Alhassane ZARE

Towards effective management of the invasive plant Senna obtufifolia

Alhassane ZARE

Towards effective management of the invasive plant Senna obtufifolia

a new threat to West African pastures

ScienciaScripts

Imprint
Any brand names and product names mentioned in this book are subject to trademark, brand or patent protection and are trademarks or registered trademarks of their respective holders. The use of brand names, product names, common names, trade names, product descriptions etc. even without a particular marking in this work is in no way to be construed to mean that such names may be regarded as unrestricted in respect of trademark and brand protection legislation and could thus be used by anyone.

Cover image: www.ingimage.com

This book is a translation from the original published under ISBN 978-620-6-72219-9.

Publisher:
Sciencia Scripts
is a trademark of
Dodo Books Indian Ocean Ltd. and OmniScriptum S.R.L publishing group

120 High Road, East Finchley, London, N2 9ED, United Kingdom
Str. Armeneasca 28/1, office 1, Chisinau MD-2012, Republic of Moldova, Europe
Printed at: see last page
ISBN: 978-620-8-18544-2

Dedication

To my father, the late ZARE Souleymane

To my mother ZARE Doko née NOMBRÉ

To my dear wife KONATE Awa

To our children ZARE Nassira and Gouffrane

I dedicate this work.

ACKNOWLEDGEMENTS

This thesis was carried out as part of the *Senna Obtusifolia* Management (S.O.M) research project entitled "**Towards an efficient management of the invasive plant *Senna obtusifolia*, a novel threat to West Africa's rangelands (Ref. 92861)" for the "Livelihood, Management, Reforms and Processes of Structural Change" program.** It was fully funded by the Volkswagen Foundation (Germany).

At the conclusion of our work, we would like to express our deep gratitude to this foundation for its financial support, which made a considerable contribution to its realization. We would also like to express our sincere thanks to all the individuals and legal entities whose invaluable contributions made this thesis possible. We are particularly grateful to

- to **Professor Oumarou OUÉDRAOGO,** for directing this scientific work. You recruited us to the SOM project team and made yourself available to supervise us, despite your busy schedule. Your guidance, humanism, scientific rigor and advice have guided us in the completion of this thesis. Our words of gratitude to you remain insufficient;

- **Professor Joseph Issaka BOUSSIM**, Head of the Laboratoire de Biologie et Ecologie Végétales (La.B.E.V.), for accepting us into his laboratory and for his invaluable advice and encouragement;

- to **Professor Adjima THIOMBIANO**, we would like to express our gratitude for the advice we received from him, mainly during the thesis points for the improvement of the scientific quality of this thesis;

- to **Dr. Moussa OUEDRAOGO**, former General Manager of the Centre National de Semences Forestières, for having allowed us to carry out our fieldwork at the Centre National de Semences Forestières, we are grateful to him;

- **Dr. Sié KAMBOU**, our internship supervisor, for his technical support and all the work he carried out during our experimental work at the Centre National des Semences Forestières (CNSF);

- **Mrs. Edith DABOUE**, current General Manager of the Centre National de Semences Forestières (CNSF), for her support and advice during our experimental work at this center;

- to the two mastorant students **Basnewindé Ilboudo, Christian Bougma, Samira SANE** and to the technician **Cyrille Sinaré**, for having accompanied us in the data collection;

- **Dr Karim OUEDRAOGO**, doctor in sociology and beneficiary of the SOM project, for his frank collaboration;

- We are grateful to **Professor Amadé OUÉDRAOGO** for his advice and encouragement;

- to **Professor Salifou TRAORÉ**, we would like to express our gratitude for his advice;

- to **Doctor Frédéric ZONGO**, whose many valuable suggestions and advice encouraged us to complete this thesis;

- **Dr Loyapin BONDE**, for his commitment and determination to support us with advice, suggestions, data analysis and encouragement;

- **Dr. Innocent Charles Emmanuel TRAORE** for his assistance with data collection and analysis. We sincerely thank him for his time and collaboration;

- **Doctor Elycée TINDANO,** for his advice, suggestions and encouragement;

- **Dr Issouf ZERBO** for his assistance with data analysis. We sincerely thank him for his time ;

- to **Doctors Blandine Marie Ivette NACOULMA, Phillipe BAYEN, Kangbéni DIMOBÉ, Moussa GANAMÉ and Mohamed CISSÉ, Emmanuel SAMBO, Blaise KABRE** we thank them for the advice and encouragement we received from them;

- **Doctors Elise SANON** *and* **Ali BOUGOUMA**, for their advice and encouragement; **Adama ZOUNGRANA** and **Hassane OUEDRAOGO** for their assistance with data analysis. We sincerely thank them for their time;

- to all our comrades of the La.B.E.V., **Aboubacar Oumar ZON, Aboubacar ZOURE, Zeinabou DABRE, Issaka KANAZOE, Séraphin HIEN, Prospère SABO for** the collaboration during all the doctoral training;

- to the **Fondation de Solidarité et d'Aide au Peuple Africain (FOSAPA)** and its Chairman **Remzi Secker** and Vice-Chairman **El hadj Noufou SAWADOGO** for their multi-faceted advice and support;

-To the **Association of Muslim Pupils and Students in Burkina Faso (AEEMB)** for its advice and support in many ways;

-to the Société d'Etude de Recherche et de Formation (**SERF**) and its first managers **Abdoulaye TARNAGDA** and **Adama ZARE** for their advice and multiform support;

I would like to thank my entire family for their support from our first steps in education right up to the present day. A special thought to the memory of my late father **ZARE Souleymane** who could not see the completion of this thesis. May the Almighty approve your works.

- To our field guides, **Ibrahim DIALLO** and **Oumboini Sosthène KOULDIATI** in DORI, to Mr **Moussa DIALLO** veterinary agent in Sidéradougou;

- To the various populations of the localities in our study area for their welcome and collaboration, which enabled us to carry out our fieldwork.

I would like to express my deep gratitude to all those who, in one way or another, contributed to the production of this document, whose names could not be mentioned in this section. To all we say THANK YOU.

Table of contents

Summary

Invasive plants modify the composition and structure of ecosystems, leading to the disappearance of many native species. In West Africa, and particularly in Burkina Faso, *Senna obtusifolia* is one of these invasive plants whose invasion has led to a reduction in natural forage, which remains the main source of food for livestock. The overall aim of this study is to contribute to better management of pastoral areas through effective control of *S. obtusifolia*. It therefore aims to propose potential solutions to control the invasion of this species. To understand people's endogenous knowledge of *S. obtusifolia*, ethnobotanical surveys targeting 300 people were carried out in villages bordering pastoral areas. This was followed by phytosociological inventories and biomass assessments of herbaceous species. A greenhouse experiment under water stress was carried out to test the competitiveness of *S. obtusifolia* against three other herbaceous species: two grasses (*Andropogon gayanus* and *Pennisetum pedicellatum*) and a legume *(Chamaecrista mimosoides)*. Finally, permanent plots were set up in a protected environment for four years to monitor the expansion dynamics of *S. obtusifolia*. Surveys were used to assess local people's perceptions of the proliferation and usefulness of this species. It emerged that the recent invasion of *S. obtusifolia* is mainly due to overgrazing. To make up for the shortage of fodder, local people harvest and store herbaceous plants and crop residues. These populations use *S. obtusifolia* leaves for food. As for the dry stem, it is used exclusively in the Sahel to make seccos. In the study area, climate, land use and topography and their interactions are the ecological factors influencing the proliferation of *S. obtusifolia*. A total of 241 plant species were recorded in 160 surveys, each covering an area of 100 m^2 . *S. obtusifolia* density was found to be higher in the Sahelian zone than in the Sudanian zone. As a result of the invasion, species richness and distribution are negatively affected. In addition, the invasion of *S.obtusifolia* considerably reduces the biomass of other species in the herbaceous stratum, from the lowest invasion level A (2.34 ± 1.09 t/ha) to the highest level C (0.82 ± 0.52 t/ha), and alters the quality of pastoral resources. Greenhouse experiments under water stress revealed that *Chamaecrista mimosoides* has a high germination rate (68.33%), followed by *S. obtusifolia* (58.66%). However, *S. obtusifolia* is the most sensitive to severe water deficit, and less competitive with *Andropogon gayanus* and *Pennisetum pedicellatum*. In the activity monitoring the dynamics of *S. obtusifolia*, the effect of the duration of protection led to a decrease in the density, biomass and height of *S. obtusifolia* at all invasion levels, while an increase in the cover rate and biomass of the other herbaceous species. Thus, in a less disturbed environment, *S. obtusifolia* regresses in terms of density and could disappear with the duration of protection. Apart from its main role of restoring vegetation and degraded soils, fencing constitutes a solution for controlling the expansion of *S. obtusifolia*. In addition, valorization of *S. obtusifolia* could also help control its expansion into pastures by popularizing its use, especially for food and handicrafts.

Keywords: invasive species; *Senna obtusifolia*; invasion factors; pastoral zone; Burkina Faso

Senna obtusifolia (L.) H. S. Irwin & Barneby proliferation in rangelands of Burkina Faso: Local knowledge, invasion factors and effects on ecosystem services

Abstract

Invasive plants affect the composition and structure of ecosystems, leading to the disappearance of native species. In West Africa, and particularly in Burkina Faso, *Senna obtusifolia* is one of these invasive plants that has led to a decline in natural fodder, which remains the main source of food for livestock. The aim of this study is to contribute to a better management of pastoral zones through an efficient control of *S. obtusifolia*. It therefore intends to propose possible solutions to control the invasion of this species. In order to understand the endogenous knowledge of people on *S. obtusifolia*, ethnobotanical surveys targeting 300 people were conducted in the villages surrounding the pastoral zones. Subsequently, a phytosociological inventory was carried out in conjunction with an assessment of the biomass of the herbaceous species. A greenhouse experiment under water stress was carried out to assess the competitiveness of *S. obtusifolia* and three other herbaceous species, including two grasses, *Andropogon gayanus* and *Pennisetum pedicellatum*, and a legume, *Chamaecrista mimosoides*. Finally, to monitor the dynamics of *S. obtusifolia* expansion, permanent plots were established in a protected environment for four years. Surveys were carried out to assess local people's perceptions with regard to the spread and benefits of this species. Overgrazing was found to be the main driver of the recent invasion of *S. obtusifolia*. To compensate for the lack of fodder, local people harvest and store herbaceous plants and crop residues. These people use the leaves of *S. obtusifolia* as food. As for the dry stem, it is used exclusively in the Sahel to make seccos. In the study area, climate, land-use patterns and topography, and their interactions, are the ecological factors that influence the spread of *S. obtusifolia*. A total of 241 plant species were recorded in 160 plots of 100 m^2 each. The density of *S. obtusifolia* was found to be higher in the Sahelian zone than in Sudan. As a result of the invasion, both species richness and species distribution are negatively affected. Furthermore, *S. obtusifolia* invasion significantly reduces the biomass of other species in the herbaceous community from the lowest invasion level A (2.34 ± 1.09 t/ha) to the highest-level C (0.82 ± 0.52 t/ha) and affects the quality of grazing resources. The greenhouse experiment under water stress showed that *Chamaecrista mimosoides* has a high germination rate (68.33%). This is followed by *S. obtusifolia* (58.66%). However, *S. obtusifolia* is the most sensitive to severe water deficit and also less competitive with *Andropogon gayanus* and *Pennisetum pedicellatum*. In the activity monitoring the dynamics of *S. obtusifolia*, the effect of protection led to a decrease in density, biomass and height of *S. obtusifolia* at all invasion levels, while the cover rate and biomass of other grasses increased. Thus, *S. obtusifolia* may decrease in density and disappear with time in a less disturbed environment. In addition to its main role of restoring the vegetation and the degraded soils, the enclosure is also a solution to control the spread of *S. obtusifolia*. Furthermore, by promoting the use of *S. obtusifolia*, especially in the food and handicraft sectors, the valorisation of this species could also contribute to controlling its expansion in the rangelands.

Keywords: invasive species, *Senna obtusifolia*, invasion drivers, pastoral zone, Burkina Faso

GENERAL INTRODUCTION

General introduction

Human actions in the form of various pressures on land (agriculture, urbanization) and the effects of climate change contribute enormously to the establishment and proliferation of invasive species (D'Antonio *et al.*, 1999; Vilà & Ibáñez, 2011; IPBES, 2019). Internationally, a large number of species are endangered, while the number of invasive species has been growing steadily over recent decades (Seebens *et al.*, 2017; Díaz *et al.*, 2019; IPBES, 2019). These invasive plants have a negative impact on biodiversity and ecosystems (Maxwell *et al.*, 2016; Fried, 2019) and are even considered to be the second most important cause of biodiversity erosion worldwide, after habitat destruction, in the endangerment and extinction of species (IUCN, 2004; Lowe *et al.*, 2007). They have severely reduced the diversity of native flora and fauna, as well as the productivity of ecosystems (Reid *et al.*, 2009). In West Africa, these invasive plants have contributed to reduced agricultural (Le Bourgeois, 2008; Thiombiano *et al.*, 2009; Pratt *et al.*, 2017) and pastoral (Bridgewater *et al.*, 2011) production, accentuating low agricultural yields and forage scarcity (Ouedraogo *et al.*, 2021). The spread of invasive species remains a global concern (IPBES, 2019) and in Burkina Faso, efforts are being made by government authorities to find solutions to this species invasion problem.

Generally, the methods used to control invasive plants are cultural, mechanical, biological or chemical (Chaves Neto *et al.*, 2020). Of all these methods, the use of chemical herbicides remains the most commonly used to control these invasive plants despite the collateral consequences on human health and soil quality (Jabran *et al.*, 2015; Nicolopoulou-Stamati *et al.*, 2016; Chauvel, 2019). Despite the widespread use and effectiveness of herbicides in controlling invasive plants, in recent years some of these plants have developed resistance to the action of certain chemical molecules (Jabran *et al.*, 2015, Meuller-Steover *et al.*, 2016). This situation makes the fight against these invasive plants even more complex. Thus, preventing these plants from infesting new areas remains more cost-effective and efficient than their extirpation (Davies and Sheley, 2007), which remains highly complex and economically less profitable (Diagne *et al.*, 2021).

West African savannas provide multiple ecosystem goods and services (Schmidt *et al.*, 2011; Ouédraogo et *al.*, 2014; Leßmeister *et al.*, 2019). While much research is devoted to the impact of invasive woody species on savannah ecosystem functions and services (Smit, 2004; Wiegand *et al.*, 2006), the role of invasive herbaceous species is less documented, particularly in West Africa. Today, Sahelian and Sudanian pastoral zones in Burkina Faso are being invaded by *Senna obtusifolia*, an annual leguminous herb native to tropical Africa and India. *Senna obtusifolia* is considered an invasive weed in South America, tropical Asia and northern Australia, where it has negative impacts on agricultural productivity and contributes to changes in ecosystem structure and function (Voss and Brennecke, 1991; Dunlop *et al.*, 2006; Sarkar *et al.*, 2012). The invasion of *S. obtusifolia* therefore poses a threat to both grazing land and livestock production. In Burkina Faso, livestock farming is one of the main sources of employment (FAO, 2019). As a result, it occupies a large proportion of the population, earns families substantial foreign currency and makes a substantial contribution to the national economy, accounting for between 10% and 20% of the country's gross domestic product (GDP). This type of livestock farming is extensive, with livestock feeding based essentially on natural pasture, which is increasingly being invaded by *S. obtusifolia*.

This species is difficult to control because it is tolerant to many commonly used herbicides (Teem *et al.*, 1974; Pitelli & Amorim, 2003). What's more, it can grow in a wide temperature range, from 32° to over 39° (Teem *et al.,* 1974, 1980; Wright *et al.,* 1999). It can even germinate on relatively dry soils (Sy *et al.,* 2001; Anastasia *et al.,* 2014; Saidou *et al.,* 2015). *S. obtusifolia* is a leguminous plant whose roots do not carry nitrogen-fixing bacteria (Boyette 2006; Anastasia *et al.,* 2014), but still manages to grow on several types of soil (Hilty, 2018) while producing a large quantity of seeds with a high germination rate (Ratzinger, 1984). It is a fast-growing plant (Tordoni *et al.,* 2019), a characteristic that gives it a competitive advantage over its congeners. In favourable habitats, *S. obtusifolia* is capable of forming dense monospecific stands and consequently provoking significant ecological changes (Mackey *et al.*, 1997; Tye *et al.,* 2003). In addition, the green leaves of *S. obtusifolia* are toxic to livestock (Burrows & Tyrl, 2006; Augustine *et al.,* 2020). Thus, its competitive ability and toxicity to domestic animals cause an ecological and socio-economic problem in agriculture and livestock farming (Walker & Oliver, 2008, Peres *et al.,* 2010). As a weed species, it has become a serious problem in soybean production throughout the southern USA (Abbott *et al.,* 1998).

Unlike on other continents, *S. obtusifolia* is not perceived as a specific threat to agriculture in tropical Africa. However, as in Australia, it can also invade natural vegetation, probably by changing its structure and function (Chaves Neto *et al.*, 2020).

In Burkina Faso, studies have been carried out on the species' forage value, invasiveness and food value (Kiema, 2008; Kiema *et al.,* 2012; Kadéba *et al.,* 2015; Zerbo *et al.,* 2016; Ouédraogo *et al.,* 2021; Nacambo *et al.,* 2021). However, there is as yet no work on the causes and impact of its invasion, let alone any method of combating it. Admittedly, Burkina Faso's latest national reports on biological diversity reveal the expansion of invasive plant species (SP/CONEDD, 2010; 2014), especially in pastures. Among these species, *S. obtusifolia* appears to be the most widespread in the landscape. However, the factors favoring its dissemination in natural vegetation, and the impacts of its invasion on ecosystem services are poorly understood. This thesis is part of the *Senna obtusifolia* Management (SOM) research project, which aims to improve pasture productivity for sustainable livestock production in pastoral areas of Burkina Faso. The general objective of this research is to contribute to the sustainable management of pastoral areas through effective control of *Senna obtusifolia* expansion.

The work is organized into four complementary activities relating to the following specific objectives:

- Understand local people's perceptions of the proliferation and their approach to managing *S. obtusifolia*. This activity addresses the following research question: How do people react to the spread of *S. obtusifolia* in their environment?
- Identify the ecological factors favoring the proliferation of *S. obtusifolia*. This activity will help you answer the following research question : Isn't the proliferation of *Senna obtusifolia* due to several factors?
- Assess the impact of *S. obtusifolia* expansion on the quality and abundance of natural forage. This activity asks the following question: does the proliferation of *S. obtusifolia* have an impact on forage availability?
- Evaluate the competitive performance of *S. obtusifolia*. This activity asks the following question: How does *S. obtusifolia* manage to colonize different habitats?

This document comprises five (05) chapters, the first of which describes the study area and the target species, *S. obtusifolia*. Chapter II examines local perceptions and use values of *S. obtusifolia*. Chapter III deals with the ecological determinants of *S. obtusifolia* proliferation and its impact on pasture ecosystem services. The competitive performance of *S. obtusifolia* according to water regime is studied in chapter IV. The final chapter (V) assesses the effect of fencing on the spatio-temporal dynamics of *S. obtusifolia* in Sahelian zones.

CHAPTER I: GENERAL INFORMATION

1. Presentation of the study area

The study was carried out in two contrasting climatic zones of Burkina Faso, namely the Sahelian and Sudanian zones, which are also accessible pastoral areas where *S. obtusifolia* is observed in abundance (Figure 1). These pastoral zones have a joint decree delimiting their perimeter, management plans and specific specifications (DGEAP, 2011). In the Sahel, the site is located in Sambonaye (37500 ha). In the Sudanian zone, the chosen site is Sidéradougou (51,000 ha).

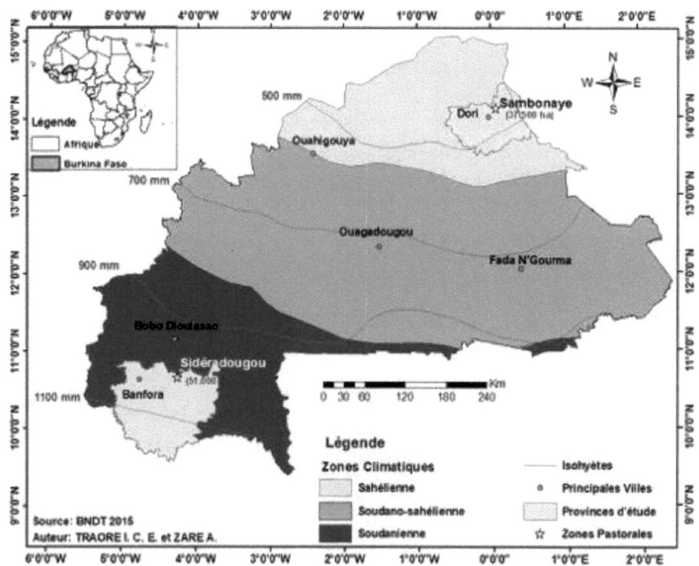

Figure 1 Location of study area

1.1. Sahelian zone

In the Sahelian zone, the study concerned the Séno province, whose capital is Dori, with an estimated population of 385900 inhabitants (INSD, 2019). In the Sahel, several socio-cultural groups cohabit: Peulh, Mossi, Gourmantché, Bella, Rimaïbé etc. (Kiema *et al.*, 2012). Agriculture and livestock breeding are the main socio-economic activities (Ayantunde *et al.*, 2020). However, the security situation is particularly worrying, with increasing violence hampering people's activities (Sirima, 2022). These populations are forced to abandon their villages, including their fields and livestock, and end up in refugee zones, swelling the ranks of social cases.

From a biophysical point of view, the soil types encountered are halomorphic, hydromorphic and sesquioxide soils, forming in places a large expanse of sand dune (IUCN, 2017). Average rainfall varies between 400 and 600 mm/year . The average annual rainfall recorded over the last thirty years (1991-2020) at the Dori synoptic station reveals a fluctuating standardized rainfall index (Figure 2). The mean annual temperature ranges from 22°C to 37°C (Figure 3). Vegetation is dominated by tree steppes composed of shrubby to thorny species including *Balanites aegyptiaca (L.) Del., Dalbergia melanoxylon* Guill. & Perr*., Vachellia nilotica* (L.) Willd. ex Del., and *Vachellia tortilis* (Forssk.) Galasso &Banfi.(Nacoulma *et al.*, 2019). The herbaceous stratum is represented by desert grasslands and maquis (Kiema *et al.,* 2014).

Herbaceous vegetation remains mainly dominated by annual grasses, and characteristic pasture grasses include *Pennisetum pedicellatum Trin., Loudetia togoensis (Pilg.) Hubb. Triumfetta pentandra* A. Rich, *Achyranthes aspera* L., *Andropogon pseudapricus* Stapf, *Aristida mutabilis* Trin, *Senna obtusifolia* (L.) H.S. Irwin & Barneby *Tephrosia pedicellata* Bak, *Acanthospermum hispidum* DC, *Schoenefeldia gracilis* Kunth; *Zornia glochidiata* C.Rchb. ex DC. (Kiema et al. 2014; Nacoulma et al 2019).

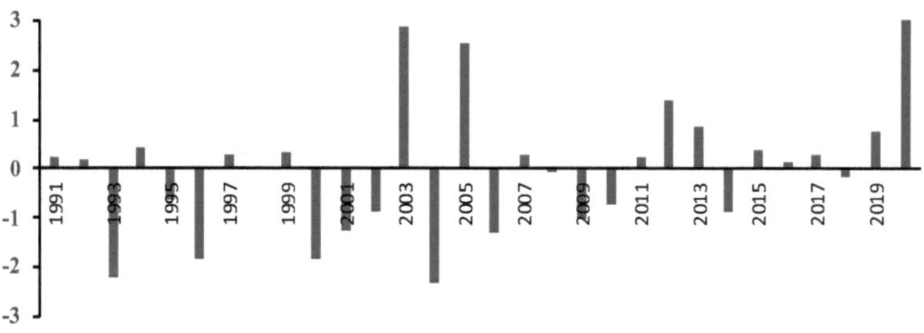

Figure 2Changes in the standardized rainfall index (SRI) in Séno province from 1991 to 2020 (Source: DGMN)

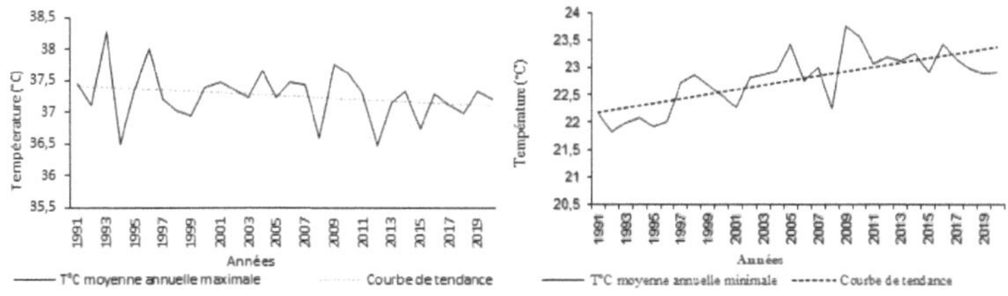

Figure 3Variation curves for mean annual maximum and minimum temperatures from 1991 to 2020 (Source: DGMN)

The study site is located in the commune of Sidéradougou situated in the Comoé province, Cascades region with an estimated population of 669,532 inhabitants (INSD, 2019). Sidéradougou is populated by Dioula, Tièfo, Dogossin, Mossi, Peulh, Bissa, Bobo, etc. (Traoré, 2011). The population's income comes mainly from agriculture and livestock (Sieza *et al.*, 2019). Soils are of the erosional lithosol type on ferruginous cuirass, retic erosion soil with ferruginous facies, eutrophic tropical vertic brown soils, gravel soils, hydromorphic soils, vertisols, granite outcrop soils (BUNASOLS, 1997; MED, 2005; Traoré *et al.*, 2020).

In the Sudanian zone, annual rainfall varies from 800 to 1100 mm, with average temperatures ranging from 24°C to 30°C. The average annual rainfall recorded over the last thirty years at the Sidéradougou synoptic station shows a variation in the amount of rainfall (Figure 4). The average annual temperature ranges from 21°C to 33°C (Figure 5).

The natural vegetation consists of savannahs, open forests with *Terminalia macroptera Guill. & Perr., Isoberlinia doka Craib & Stapf* and *Pterocarpus erinaceus* Poir. and gallery forests with *Cola laurifolia* Mast. and *Vitex chrysocarpa* Planch. ex Benth.(Nacoulma *et al.*, 2019).

Herbaceous vegetation is dominated by perennial grasses and typical rangeland herbaceous species such *as Schizachyrium platyphyllum* Stapf, *Andropogon africanus* Franch, *Acroceras amplectens* Stapf., *Hyparrhenia rufa* (Nees)Stapf, *Brachiaria distichophylla* (Trin) Stapf, *Paspalum scrobiculatum* L., *Panicum phragmitoides* Stapf, *Hyptis suaveolens* Poit, *Sida urens* L., *Senna obtusifolia* L. H.S. Irwin & Barneby (Kiema *et al.*, 2014).

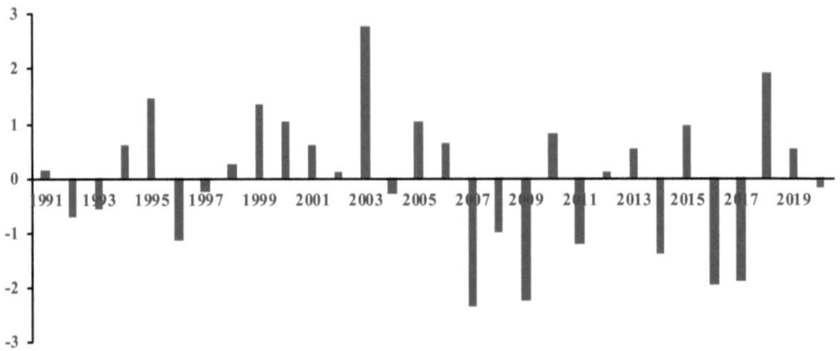

Figure 4Change in standardized rainfall index (SRI) in Kénédougou province from 1991 to 2020 (Source: DGMN)

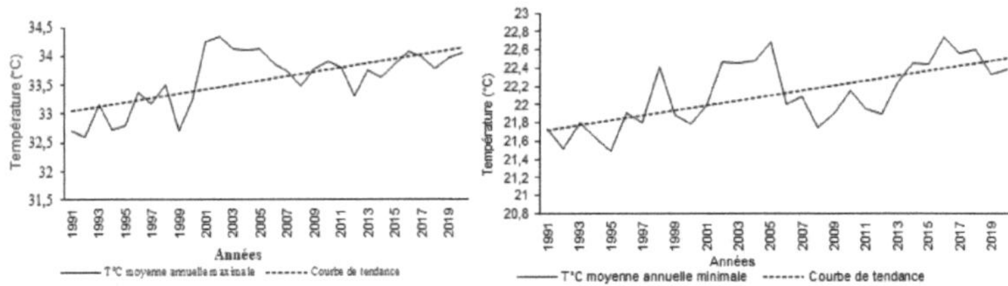

Figure 5Variation curves for mean annual maximum and minimum temperatures from 1991 to 2020 (Source: DGMN)

2. Pastoral zone

Pastoral zones are specially developed pastoral areas whose main purpose is to carry out pastoral activities (DGEAP, 2011). They comprise subdivided terroir areas, the first part of which is reserved for grazing, the second for forage crops and the third for herders' dwellings. Pastoral zones are identified as such by national, regional, provincial and land-use planning schemes, and earmarked for pastoral development operations.

2.1. Justification for the creation of pastoral zones

In West Africa, and particularly in Burkina Faso, extensive livestock farming is the most common system, requiring herds to be mobile in search of food. Pastoral livestock farming concerns more than 80% of domestic ruminants (cattle, sheep, goats) in Burkina Faso (Ouoba-Ima, 2018). Animal feed is based mainly on natural pastures, post-cultivation pastures and salt cures. In the face of galloping demographics, leading to an increase in cultivated areas to the detriment of pastures, and in the face of a significant increase in the number of ruminant livestock (Obulbiga *et al.*, 2018), a forage imbalance has emerged that tends to increase from year to year. This situation makes transhumance more difficult and aggravates conflicts between herders and other users of natural resources. These conflicts result in the slaughter of animals, the destruction of property, land withdrawals, loss of human life, and the departure of breeders, particularly cattle breeders, from the North to the southern parts of the country and to coastal countries, where some settle despite the difficulties they encounter there and the expulsions to which they may be subjected (DGEAP, 2011). All the above constraints have led to a start being made on developing, securing and enhancing pastoral areas, particularly those affected. The objectives assigned to pastoral areas are to :

- long-term sedentarization of livestock farming;

- organize breeders and intensify livestock production for the market;

- integrate agriculture and livestock farming and produce draught oxen ;

- combat desertification and manage natural resources rationally ;

- ensure concerted management of space and natural resources ;

- secure land tenure for livestock and livestock production.

2.2 Spatial organization of pastoral zones

In each pastoral zone, a minimum of 80% of the area is reserved for grazing. The remaining 20% (housing area) is occupied by the farmers. Housing areas are subdivided into "fermettes". Each fermette comprises five (5) different areas assigned to the farmer's home, forage crops, food crops, reforestation and fallow land (DGEAP, 2011).

2.3 Institutional framework governing pastoral zones

At institutional level, the pastoral zones came under the supervision of the initiating structures. Today, they are attached to the Directions Régionales des Ressources Animales (DRRA) of their territorial jurisdiction. At national level, the Direction

Générale des Espaces et des Aménagements Pastoraux (DGEAP, 2011) is the structure responsible for pastoral development.

2.4. Current situation of pastoral zones

There are twenty-six (26) developed pastoral zones and 156 potential pastoral zones and grazing areas. They cover an area of around 2,000,000 ha (DGEAP, 2011).

There are 26 functional pastoral zones, covering a total area of around 730,960 ha. Most of them have a joint decree delimiting their perimeter, management plans and specific specifications (DGEAP, 2011). This is the case of the Sidéradougou pastoral zones in the Sudanian zone and the Sambonaye and Ceekol Nagge pastoral zones in the Sahelian zone, which constitute the zones of this study. These three (03) functional pastoral zones have management master plans, two (02) of which have been validated at local level (Sidéradougou and Sondré - East) and one (01) at national level (Nouhao). However, the areas initially identified tend to shrink enormously, as in the case of the Sidéradougou pastoral zone, which has shrunk from 307,000 ha to around 51,000 ha today (DGEAP, 2011). This reduction is due to the occupation of pastoral areas by farmers for agriculture and housing construction (Photo 1 & Photo 2). It should also be noted that since 2002, the crisis in Côte d'Ivoire has led to a massive return of returnees to the area (Ouédraogo and Thiombiano, 2017). These ever-growing populations have monopolized the land. This has resulted in the extension of cultivated areas and the development of extensive livestock farming, encroaching on forest and pastoral resources.

Much of the land allocated to herders has been cultivated or reclaimed for other activities, including farming and plantations (Higazi and Aboubakar, 2018). High demographics, combined with unfavorable environmental and climatic conditions, have also reduced the amount of usable land. There is a system of transhumance routes via water points, rivers or boreholes, but many of these have had their access blocked or reduced by crop fields (Higazi and Aboubakar, 2018). Pressure on pastoral livelihoods and farmers has therefore increased. As for pastoral areas in the Sahel, they are being degraded and some are being invaded by *S. obtusifolia*, a species of low forage value (Kiéma, 2012). This contributes to reducing the quality and quantity of forage in these areas.

Photo 1Boundary marker for the Sidéradougou pastoral zone
with Sorghum field illegally inside

Photo 2Illegal settlements in the Sidéradougou pastoral zone

3. Introducing *Senna obtusifolia*

3.1 History

Initially considered to be *Cassia obtusifolia L.* on the basis of fruit structure, Bentham (1871) argued that *C. obtusifolia* and *C. tora* were the same species and suggested that they should both be referred to as *C. tora*. Thus, some publications claim that *C. obtusifolia* is synonymous with *C. tora* (Holm *et al.*, 1979, 1997; Irwin and Barneby,

1982; Abbot *et al.,* 1998), while others distinguish them as different species (Linnaeus, 1753; De Wit, 1955; Brenan, 1958; Randell, 1995). Since then, there has been confusion in the literature and debate among taxonomists as to whether *C. obtusifolia* and *C. tora* are the same species or independent species (Crawford *et al.*, 1990).

C. obtusifolia was subsequently placed in the genus *Senna* by Irwin and Barneby in 1982, underlining the importance of floral morphology for classification (Marazzi *et al.*, 2006). However, they did not include *Cassia tora* L. (a species not found in America) in their revision, as Irwin and Barneby (1982) focused solely on species found in the South American collection, considering the two species synonymous (Marazzi *et al.*, 2006). In 1988, when Randell used the three genera to replace the genus *Cassia* in the broad sense, he placed *Cassia tora* in the genus *Senna*, hence *Senna tora* (L.) Roxb. Today, in addition to the Cassia genus, there is *also* the Senna genus and the Chamaecrista genus. All the species grouped in the genus Cassia in the broad sense have been divided between these three genera, so that the genus *Cassia* in the strict sense comprises 30 species, *Senna* 260 species and *Chamaecrista* 270 species (Grubben and Denton, 2004).

3.2. Description and ecology

Senna obtusifolia (L.) Irwin & Barneby, is a therophyte of the kingdom plantae, phylum spermatophyta, class dicotyledons and belonging to the order Fabales, family Fabaceae, subfamily Caesalpinioideae (Binggeli, 2005; Thiombiano, 2008). *S. obtusifolia* is derived from the Latin words obtusus, meaning obtuse or blunt, and folium, meaning leaf. Both refer to the shape of the leaf. *S. obtusifolia* in English is sicklepod, in French Séné or pistache marron. Its name in Mooré is "Katr-nangouri" for some and "Sôgda" for others. In Fulfulde, it's called "Houlô". Among the Dioula, the species is called "kri-kri". *S. obtusifolia* leaves are alternate, imparipinnate, with 3 pairs of leaflets (photo 3). Stipules are linear or filiform. Leaflets are obovate, 1.5-5(-6) cm long. The inflorescence has a generally very short peduncle, 1-2 flowers, bisexual, 5-mothered. The pedicel is long, the sepals oval, about 5 mm long and the petals obovate, 1-2 cm long, yellow, 10 stamens. The ovary is superect, linear and curved. The fruit is a linear, dehiscent pod, up to 23 cm x 0.5 cm, straight or curved, with numerous rhomboid seeds, about 5 mm long. *S. obtusifolia* multiplies by seed (Ratzinger, 1984). Optimum conditions for seed germination of this species are temperatures between 15 and 30°C, pH between 5 and 7, with maximum germination at pH = 6.1 (Norsworthy *et al.*, 2006). The seedling has epigeous germination and two semi-fleshy cotyledons (Grubben and Denton, 2004). It is a weed of plateaus, field margins and roadsides, particularly in forested areas (Jonhson, 1997). *S. obtusifolia* is also found along rivers and lake shores, and on cultivated land up to 1700 m altitude (Grubben and Denton, 2004). It is a ruderal species, highly disseminated by livestock (Thiombiano *et al.*, 2012). The few distinguishing features between the two species are: regarding the leaf gland, S. *obtusifolia* has 1-2 glands on the two lower pairs of leaflets. The second gland, when present, is only on the lower leaflets (Brenan, 1967). However, S. tora has 2 glands on both pairs of leaflets. The second gland is located only between the middle pair of leaflets (Brenan, 967). Phytochemically, *S. obtusifolia* contains dihydroxy anthroquinones, obtusin and obtusifolin (Upadhyaya and Singh, 1986). *In S. tora*, monohydroxy anthraquinone, chryso-obtusin (Upadhyaya and Singh, 1986).

Photo 3 Adult plant of *Senna obtusifolia*

3.3 Origin and distribution

The original range of *S. obtusifolia* was limited to tropical South America, but it has spread to all tropical and subtropical regions of the world (Rendell, 1995). However, its actual range is poorly known, as *Senna obtusifolia* is often confused with *Senna tora*, an Asian species found in India, China, the Philippines and other parts of Asia (Brenan, 1967; Randell, 1995). In some regions, such as India and South Asia, both species can be found cohabiting in the same environment (Rendell, 1995). Thus, American and European researchers familiar only with *S. obtusifolia* have considered the names of the two species to be synonymous (Rendell, 1995). By contrast, Asian researchers familiar with both species were able to differentiate between them (Rendell, 1995).

Today, *S. obtusifolia* has a pantropical distribution (Mackey *et al.*, 2014). It is found throughout tropical Africa, with the exception of Madagascar, in Asia and the Philippines. It is thought to have been introduced early to Africa from America. African plants are thought to be of Caribbean origin, with large fruits like those of specimens from the Caribbean and southern USA (Rendell, 1995). *S. obtusifolia* is considered a weed worldwide, with an estimated 600,000 hectares infested in Queensland, Australia. The presence of *Senna tora* in Africa is doubtful, and when it is mentioned, it is probably *S. obtusifolia* (Grubben and Denton, 2004). On a regional scale, *S. obtusifolia* has a Sahelian to Guinean distribution. According to the list published by Binggeli in 2004, *Senna obtusifolia* has been confirmed in Africa in countries such as Benin, Ghana, Mali and Niger (Holm *et al.,* 1979). In Burkina Faso, the species has been sampled in all climatic zones, with a high probability of occurrence in Sudanian to Sudano-Sahelian zones (Thiombiano and Kampmann, 2010).

3.4. Uses

The leaves of *S. obtusifolia* are the most widely used parts of the species. They are used as vegetables in many countries across Africa and India (Pasternak *et al.*, 2007). In Sudan, the leaves are fermented to produce 'kawal' (a condiment used in sauce), which is consumed and sold (Abakar *et al.*, 2019). The seeds are used as food in times of famine (Grubben and Denton, 2004). In traditional medicine, *S. obtusifolia* seeds are used against venomous insect bites (Eklu-Natey and Balet, 2012). Coffee made from the seeds is a remedy for stomach aches and fatigue (Dirar, 1984).

Consumption of *S. obtusifolia* by cattle, sheep, goats and ostriches depends on its stage of development. In cattle, *S. obtusifolia* can cause mycotoxicosis, an often fatal disease probably caused by mycotoxins, which are produced by the fungi affecting *S. obtusifolia* (Grubben and Denton, 2004). Leaves, roots and seeds are used for gastric, eye and body care. The stems are used to make mats and fences. *S. obtusifolia* is often used as an ornamental plant (Grubben and Denton, 2004). In India, gums extracted from *S. obtusifolia* seeds are used in the food industry. Extracts of *S. obtusifolia* are used or tested as pesticides (Grubben and Denton, 2004; Thiombiano *et al.*, 2012).

4. Biological invasion concepts and definitions

4.1 Native species

An autochthonous or native species is one whose presence in a region is the result of natural processes without human intervention (Pyšek *et al.* 2004; Essl *et al.*, 2018). It is native to a given territory in relation to a reference period (Thévenot, 2013; Yohann, 2015).

4.2 Allochthonous species

An allochthonous or exotic species is a species introduced (voluntarily or accidentally) by man into a new territory outside its natural range (Richardson *et al.*, 2000). According to Pereyra (2013), a foreign or non-native or exotic species refers to a species outside its current or historical native range. The vast majority of introduced species are not and are unlikely to be invasive (Yohann, 2015). These species are known as naturalized plants.

4.3 Naturalized species

A naturalized plant species is a foreign plant that has been introduced fortuitously or voluntarily, and which reproduces by seed or vegetatively on a regular basis over several life cycles, and which has extended its area of occurrence significantly since its introduction. It does not necessarily invade its environment (Richardson *et al.*, 2000).

Depending on the extent of naturalization, a distinction is made between naturalized amphibians for species that are widely naturalized on a large scale and spread rapidly by mixing with the native flora, and stenonaturalized amphibians for locally naturalized plants (Nzigidahera, 2017).

4.4 Cryptogenic or cryptogenic species

Refers to a plant for which it is impossible to determine, after evaluation, whether it is indigenous or exotic. For species whose era of origin is unknown, the expression "cryptogenic species" or "cryptogenic species" should be used (Thévenot, 2013).

4.5. Biological invasion

There are many definitions of what constitutes a biological invasion, and not all of them describe exactly the same process (Gard, 2012). The main points of divergence concern the geographical criterion and the impact criterion (Valéry *et al.*, 2008). The geographical criterion means that biological species have been introduced (allochthonous) by man across biogeographical barriers (Richardson et *al.*, 2000). This definition categorically excludes native species, which can also become invasive (Valéry *et al.*, 2008). The impact criterion refers to the consequence of biological invasion on ecosystems (Tassin *et al.*, 2017). It is therefore to take these two criteria into account that Black and Bartlett (2020) emphasize that invasive organisms refer to native or exotic species that spread, with or without human assistance, in natural or semi-natural habitats, producing a significant change in ecosystem composition, structure or processes, or causing serious economic losses for human activities.

In English, the term 'invasive species' is used to describe invasive species. In French, however, the two words invasive and envahissante seem to have different meanings. The difference between these two terms is the application of the geographical criterion (Gard, 2012). Thus :

- An **invasive species** is equivalent to an alien invasive species. It has been introduced into an environment where it has naturalized and become prolific, further extending its geographical distribution and transforming or is in the process of transforming biodiversity (Thévenot, 2013). According to Yohann (2015); Roussel and Blackburn (2016), it is a species introduced voluntarily or accidentally by Man, into a new territory outside its natural range, whose establishment and spread threaten ecosystems, habitats or native species with negative consequences on ecological and/or socio-economic and/or health services.

- The acronym **"IAS"** is used to designate **"invasive alien species"**. However, according to Valéry *et al* (2008), an invasive species is any species that, following the disappearance of natural barriers to its proliferation, expresses a dominance enabling it to spread rapidly and conquer new areas in its ecosystem. This expression can be used for both native and non-native species when there is some form of similarity between environmental conditions (Pyšek *et al.*, 1995; Prach & Pyšek, 1999). Thus, any species, whether native or non-native, that multiplies rapidly, extends its range and becomes a nuisance or even a nuisance is said to be invasive (Thévenot, 2013). For the purposes of this thesis, the definition of invasive plant is best applied to *S. obtusifolia*, which is a largely naturalized (Nzigidahera, 2017), i.e. local, invasive plant.

4.6. Characteristics predisposing a species to become invasive

Native or non-native invasive species have different biological traits to non-invasive species (Valéry, 2006). Some invasive alien species appear to share common biological traits (Yohann, 2015):
- a high capacity for reproduction or multiplication;
- rapid development, making them highly competitive with other species;
- strong adaptability and resistance to disturbance;
- few or no natural predators.

4.7. Biological invasion processes

The process of biological invasion is complex and takes place in five stages: importation, introduction, acclimatization, reproduction and dispersal-colonization (Richardson *et al.*, 2000; Williamson and Fitter, 1996; IUCN, 2015). The passage from one stage to another requires the crossing of one or more natural barriers, with or without human assistance. During importation, the species leaves its territory of origin to find itself in a new territory (geographical barrier). The introduction of a species into an environment can be voluntary or involuntary (Yohann, 2015). The imported species arrives in a new territory. In this new environment, the introduced species acclimatizes to the local conditions (climatic barrier) and reproduces to reach the stage of stable populations (biological barriers through pollination). This is called naturalization. Subsequently, the naturalized species can disperse and colonize new habitats (ecological barrier). It then becomes invasive: we speak of invasive species or invasive alien species (Williamson and Fitter, 1996; Richardson *et al.*, 2000; Yohann, 2015). For plants in particular, invasiveness can become apparent decades or even more than a century and a half after the arrival of the new species on the territory: this is known as the latency phase. Not all introduced species become invasive. Out of every 1,000 species imported by man, only one will become invasive in its host environment (Figure 6). This is Williamson and Fitter's "Three tens rule" (1996). Biological invasion is therefore a long process involving the dispersal and colonization of a new area by a species that has been introduced there.

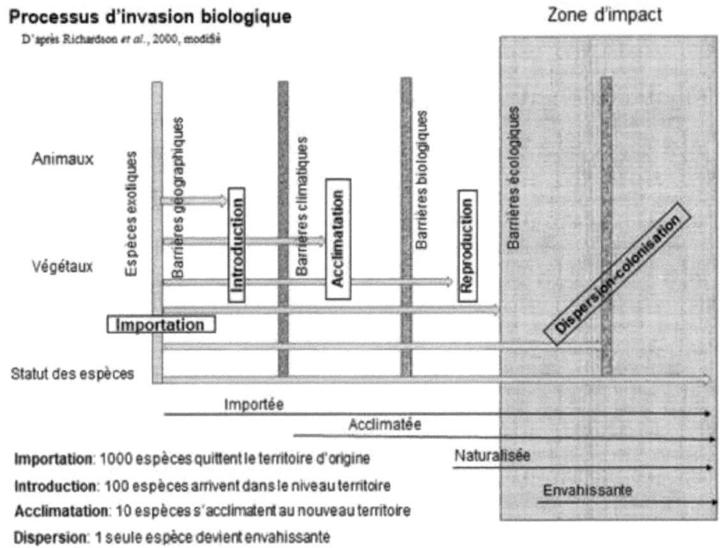

Figure 6Biological invasion process, after Richardson et *al.* (2000)

4.8. Examples of invasive species in Burkina Faso

Burkina Faso's latest national biodiversity reports reveal the expansion of "invasive" aquatic and terrestrial plant species. Among the herbaceous plants concerned are *Eichhornia crassipes* (Mart.) *Solms*, *Azolla africana* Desv., *Mimosa pigra L., Hyptis suaveolens (L.)* Poit, *Senna occidentalis* (L.) Link, *Lippia chevalieri* Moldenke, *Senna obtusifolia* (L.) H.S.Irwin & Barneby (SP/CONEDD, 2010 ; 2014).

4.9. Ecosystem services

Ecosystem services refer to the benefits that people derive from ecosystems. They fall into four (04) categories: **supporting services, provisioning services, regulating services and cultural services** (Cissé *et al.,* 2018). The quality and maintenance of human life on earth is highly dependent on these ecosystem services (Thiombiano and Kampmann, 2010). For Balay *et al.* (2015), provisioning or production services and regulating services are encountered in ecosystems arising from livestock activities. In an assessment of the main ecosystem services provided by grasslands, Puydarrieux and Devaux (2013), consider on the one hand that regulating services can generate provisioning services, and on the other that supporting or self-maintenance services are necessary for all other services. For this reason, they distinguish several ecosystem services per category in grasslands:

- regulating services: carbon fixation and storage, regulation of other gases, water and, protection against flooding and erosion, pollination, biodiversity ;
- supply services: livestock products such as fodder and plant proteins, harvested products such as flowers, berries and mushrooms;
- cultural services: hunting, landscape interest, education and scientific knowledge.

CHAPTER II: LOCAL KNOWLEDGE AND USAGE VALUES OF *SENNA OBTUSIFOLIA* (L.) H. S. Irwin & Barneby

This chapter was the subject of a publication:

Zaré Alhassane, Traoré Innocent Charles Emmanuel, Hien Bossila Séraphin, Bonde Loyapin & Ouédraogo Oumarou (2022). Local Knowledge, Perceptions and Uses Values of *Senna obtusifolia* (L.) H.S. Irwin & Barneby, an Invasive Native Plant Species in Burkina Faso, West Africa. Scientific Journal, ESJ, 18 (30), 343. https://doi.org/10.19044/esj.2022.v18n30p343.

1. Introduction

Invasive species are listed among the main threats to biodiversity conservation, leading to ecosystem modification and loss of ecosystem services (Walsh *et al.*, 2016; Kueffer, 2017; Pyšek *et al.*, 2020). As a result, many studies have focused on the detrimental effects of these invasive species on biodiversity and ecosystems (Shackleton *et al.*, 2019; Archibald *et al.*, 2020; Bailey *et al.*, 2020). However, invasive species could have positive ecological and socio-economic impacts (Charahabil *and* Akpo, 2018; Wagh *and* Jain, 2018; Souley *et al.*, 2020) that can be used as an adaptation strategy to their negative effects. For example, some of these species provide ecosystem services of provisioning (food, medicine) and support (soil fertilization) to populations (Shackleton and Shackleton, 2018). Indeed, a recent study indicated that local populations in Burkina Faso use neglected species, including invasive species, as a resilience strategy (Ouedraogo *et al.*, 2021). *Senna obtusifolia* is one of the invasive native species (SP/CONEDD 2014) that proliferates in habitats under intense grazing pressure (Kiema *et al.*, 2014). These young green leaves are used as food in some African countries (Abakar *et al.*, 2019), indicating that this invasive species is potentially valuable. Other studies have mentioned the use of *S. obtusifolia* to treat certain diseases (Cunningham,1993; Eklu-Natey & Balet, 2012). However, in Burkina Faso and at the current stage, the usefulness of this species remains little explored (Nacambo *et al.*, 2021). According to García-Llorente *et al.* (2008), local people's perceptions of invasive species and traditional forms of valuing these species could help design effective and sustainable management policies. Consequently, the objectives of this study were to:

- identify the endogenous knowledge of local populations about the invasion of *Senna obtusifolia* ;
- assess the use value of the species in both climatic zones;
- identify the uses of the species in the two climatic zones.

The assumptions underlying this activity stipulate that:

- endogenous knowledge of *Senna obtusifolia* is associated with the specific socio-cultural group of the populations ;
- the forms of use of the species are more diversified in the Sudanian zone

2. Materials and methods

The study area is that of Dori in the Sahel and Sidéradougou in the Sudanian, as described in Chapter I above.

2 .1. Sampling plan

A stratified sampling plan based on climatic zones and people's social profile was adopted. Climatic conditions are significant factors affecting the presence and abundance of plant species, as well as the availability of resources to people, and therefore could probably influence people's knowledge, perception and preferences in species use. Social profile (ethnicity, gender and age) is also an important factor affecting the management and use of local forest resources, including preferences and valuation (Ouédraogo *et al.,* 2013; Agúndez *et al.*, 2020). Ethnobotanical surveys were carried out in Dori and Sidéradougou. The stratified probability sampling or proportional stratified random sampling method was used (Masengo *et al.*, 2021). It involves dividing the study area into different strata, represented here by the villages, and considering the same number of respondents in each (Inkoto *et al.,* 2019). Interviews were conducted with the populations of ten (10) villages located near pastoral zones, at a rate of five (05) villages per climatic zone. Eight (08) socio-cultural groups were selected on the basis of their knowledge and use of the species. This selection was made on the basis of information provided by local leaders (village chiefs, heads of village associations, delegates or CVDs). In the Sahelian zone, Gourmantché, Sonrai and Bella were selected. In the Sudanian zone, Bobo, Tieffo and Dogossin were selected. The two cross-cutting groups are the Mossi and Peulh of the two climatic zones. These two groups are the most dominant in Burkina Faso (Sop *et al.,* 2012). They may have more knowledge about the perception and use of *Senna obtusifolia* as they are well distributed throughout the country. To confirm the information received by officials in the various localities, a preliminary survey was carried out on a random sample of 100 people in each study site to determine the proportion of the study sample. This survey showed that 50 people knew and used *Senna obtusifolia*. The number of informants in each area was determined by the following formula (Dagnelie 1998):

$$n = \frac{U_{1-\alpha/2}^2 P(1-P)}{d^2}$$

Where n is the sample size, p the proportion of respondents who know and use the species (p =0.5), U1-∝/2 =1.96 is the value of the reduced normal variable set **for** a probability value ∝= 0.05 and d the margin of error set at 0.08. A sample size of 150 people in each zone was determined.

2.2 Interview

As indicated in the sampling plan, surveys were conducted considering the factors climate (02 climatic zones), ethnicity (08 socio-cultural groups), gender (02 gender balance: men and women) and age (03 age classes). Each factor combination was sampled with 5 replications, giving 2x5x2x3x5 = 300 respondents interviewed. The three age classes adopted were: people aged at least 18 and with an age ≤30 years, 30 < age < 60 years and people with an age ≥ 60 years (Assogbadjo *et al.*, 2008). They were selected on the basis of their voluntary adherence and knowledge of *Senna obtusifolia*. People's socio-demographic characteristics (socio-cultural group, gender and age) are important factors influencing the management and uses of forest products, including preferences and valuation (Ouédraogo *et al.*, 2013; Agúndez *et al.*, 2020). The survey technique used was based on semi-structured interviews, using a previously tested questionnaire (Alexiades and Sheldon, 1996) to ensure that the questionnaire provided all the information sought. The main information gathered was ethnobotanical knowledge of the species, i.e. invasion factors, impact of *Senna obtusifolia* on forage availability, different uses of the species (categories of *Senna obtusifolia* use and organs used) (Appendix 1). Interviews were conducted with the help of native interpreters, and samples of *Senna obtusifolia* were collected and brought to the study sites, and presented to the informants during the interviews to ensure the presence of the species in each village.

2.4. Data analysis

Before performing statistical tests between factors, the following parameters were calculated:

- The species' ethnobotanical use value (EUV) using the equation proposed by (Phillips *et al.*, 1994): EUV=ΣUi/n, where Ui is the number of uses mentioned by informant i and n is the total number of informants;
- Use value per organ and response rate (T) using the equation proposed by (Maregesi *et al.*, 2007):

$$T = 100 * \frac{S}{N}$$

Where S is the number of people who gave a positive 'yes' response from the organ concerned, N is the total number of informants. The response rate, ranging from 0 to 100, indicates the most frequently used organs in the study area. For each ecological knowledge cited, a Chi-square test was carried out to determine the statistical differences between the climatic zones and the social profile of the informants (gender, age and socio-cultural group). The statistical value was tested at the 5% level. The use value (EUV) of *Senna obtusifolia* was analyzed via a generalized linear model (GLM) with a Poisson distribution (Zuur *et al.,* 2009) to test its variation as a function of study site and socio-demographic factors. The Student Newman Keuls (SNK) test of separation of means was used to test for differences between socio-cultural groups. All analyses were performed using software version 4.1.0 (R Core Team, 2021).

3. Results

3.1. Local names for *Senna obtusifolia* and their ethnic significance

Eight (8) local names were identified within the eight socio-cultural groups. The names vary between socio-cultural groups, with the exception of the two cross-cultural groups (Mossi and Peuhl), each of which uses the same name for the species, regardless of the climatic zone in which it is found (Table 1). Meanings differ according to socio-cultural group, and are based on taste, natural habitat, utility and resemblance to other plants (Table 1).

Table 1Local names for *Senna obtusifolia* according to socio-cultural groups

Zone	Socio-cultural group	Local name	Meaning
Sahelian	Bela	Kanaga	Edible leaf grass
	Gourmantché	Palemkpamba	Unknown
	Mossi	Sôgôda/katre nangouri	Hyena solder/peanut food
	Peulh	Oulon	Lowland grass
	Sonrai	Oul-oula	Useful grass
Sudanese	Bobo	Kikir	Bitter-tasting leaf
	Dogossin	Tiguinfara/kikir	Peanut skin
	Mossi	Sôgôda/katre nangouri	Hyena solder/peanut food
	Peulh	Oulon	Bottomland grass
	Tiéffo	Konon kassosso	The bird bean

3.2. Local perceptions of the *Senna obtusifolia* invasion and forage management strategies

In both climatic zones, informants confirmed the invasion of *S. obtusifolia* in many parts of their landscapes. They added that in the last decade, the species was mainly observed in lowlands, but nowadays it is also found on roadsides, grazing areas (rangelands and fallow land). Animals and humans are the factors most cited by informants as being responsible for the spread of *S. obtusifolia*. A significant difference between factors was observed according to zone (X^2 =27.23; df=3; p<0.0001). In the Sahelian zone, animals were the most cited factor, with a frequency of 59.33%, whereas in the Sudanian zone, it was human activities (56%) that favored the invasion of *Senna obtusifolia* (Table 2). Other factors, such as water and wind, were the least cited in both zones. Furthermore, 90% of informants recognized that the invasion of *Senna obtusifolia* in pastures had a negative impact on forage quality and abundance (Table 2). Despite this situation, 55% of informants in both climatic zones found the species useful to them (Table 2).

In the Sudanian zone, the harvesting and conservation of crop residues is the most common alternative to compensate for the lack of herbaceous forage. In the Sahelian zone, mowing herbaceous forage and harvesting crop residues are the most common practices (Table 2).

Table 2Perceptions of *Senna obtusifolia* by populations and frequency of explanatory variables by modality of the dependent variable

	Sahelian zone		Sudan zones		
Variables	Absolute frequency	Relative frequency	Absolute frequency	Relative frequency	Chi-square test
Invasion factors					
Animal	89	59,33	48	32	
Human	42	28	84	56	X^2=27,23
Water	17	11,33	14	9,33	df=3, p<0.0001*
Wind	2	1,33	4	2,66	
Impact of *S. obtusifolia* on forage value					
Negative	140	93,33	132	88	X^2=1.5102, df=1
Positive	10	6,66	18	12	p<0.0001*
Importance of *S. obtusifolia* for populations					
Important	102	68	66	44	X^2=16.57 df=1
Not important	48	32	84	56	p<0.0001*

* indicates a significant difference

3.3. *Senna obtusifolia* uses and harvested organs

Eight uses for *Senna obtusifolia* were identified (Figure 7). Food was the most important (100%) use of *S. obtusifolia*, followed by construction (49%), handicrafts (26.66%), trade (20%) and medicine (11.66%). The most widely used organs were leaves (100%) and stems (50%), as shown in Figure 8. Leaves are used for food (sauce, couscous), while stems are used for construction (house roofing), handicrafts (making seccos), wood energy and fertilization. Seeds and roots were less widely used. Seeds were used for both food and medicine, while roots were only used for medicinal purposes (Figure 8).

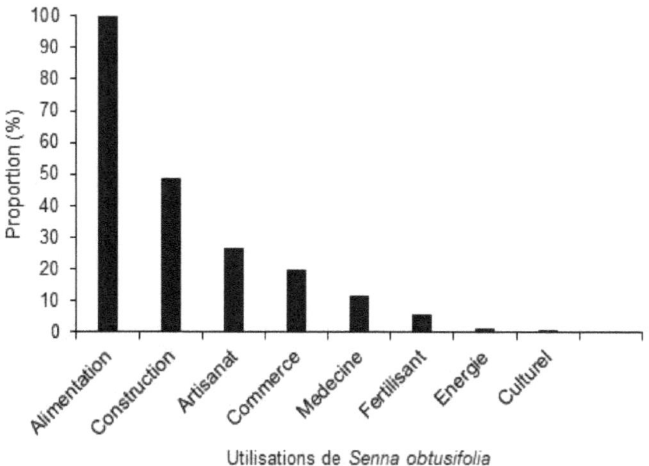

Figure 7Frequency of *Senna obtusifolia* use citations

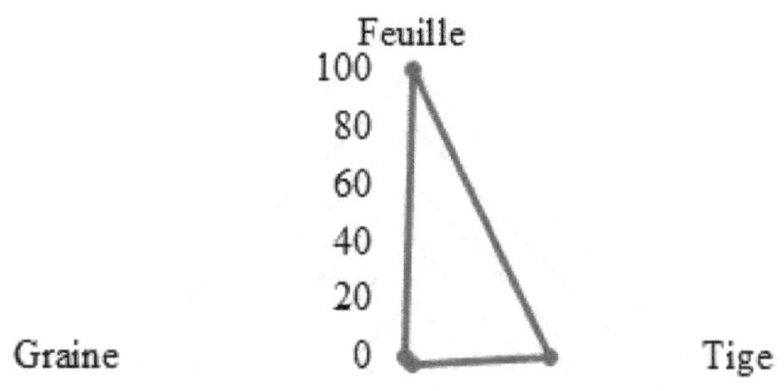

Figure 8Frequency of quotations for parts used on *Senna obtusifolia*

3.4.Ethnobotanical use value of *Senna obtusifolia* according to climatic zones and socio-demographic characteristics

The ethnobotanical use value (EUV) of *Senna obtusifolia* varied significantly (p < 0.001) between climatic zones and sociocultural groups (Table 3). Age and gender had no influence on this ethnobotanical use value.

Table 3Influences of climatic zone and socio-demographic factors on the use value (UV) of *Senna obtusifolia*

	VU	Deviation	t value	Pr(>|t|)
Factors	-2,764	0,246	-11,254	< 2e-16 ***
Climate zone	0,341	0,090	3,794	0,001 ***
Ethnicity	-0,058	0,015	-3,787	0,001 ***
Age	0,004	0,046	0,090	0.928ns
Gender	0,005	0,075	0,073	0.941ns

*** =Significant, ns= not significant

In the Sahelian zone, the Bela use the species more than any other socio-cultural group (Figure 9). In contrast, in the Sudanian zone, there was no significant difference in the use of *Senna obtusifolia* between the five ethnic groups. Furthermore, if we consider the two cross-cultural groups, there was a significant difference in the use of *Senna obtusifolia* between the Peulh and Mossi in the Sahelian and Sudanian zones (Figure 9).

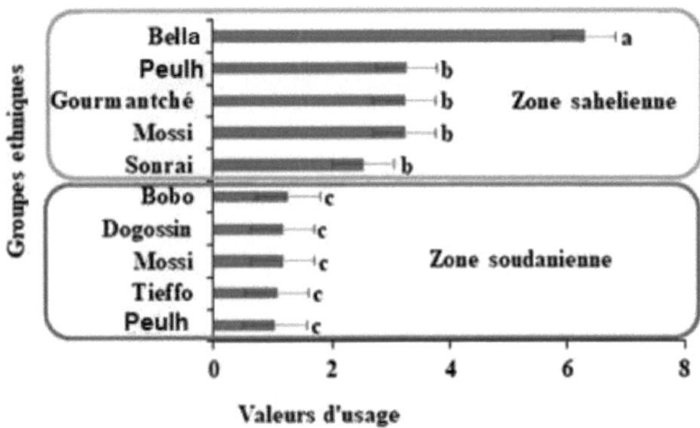

Figure 9Total use value (UV ± standard deviation) of *Senna obtusifolia* by ethnic group
Levels not linked by the same letters are significantly different

3.5.Local fodder acquisition strategies in areas invaded by *Senna obtusifolia*

Harvesting woody fodder, mowing and preserving herbaceous plants, harvesting and preserving crop residues and growing fodder crops are the methods used by the respondents to obtain fodder. In the Sahelian zone, mowing and conservation of herbaceous plants (0.76 ± 0.43) and harvesting and conservation of crop residues (0.69 ± 0.46) are the most dominant fodder production practices. In contrast, the harvesting and conservation of crop residues (0.95 ± 0.22) remains the most common forage activity in the Sudanian zone (Table 4).

Table 4Types of forage in the two zones

Types of forage	Sahel	Sudanese	W	P
Harvesting woody forage	$0,31 \pm 0,46^a$	$0,03 \pm 0,16^b$	14400	< 0,001
Mowing and grass conservation	$0,76 \pm 0,43^a$	$0,07 \pm 0,25^b$	19050	< 0,001
Harvesting and storing crop residues	$0,69 \pm 0,46^a$	$0,95 \pm 0,22^b$	8400	< 0,001
Forage farming	$0,05 \pm 0,22^a$	$0,0 \pm 0,0^b$	11850	< 0,004

Levels not connected by the same letter are statistically different

4. Discussion

4.1. Local perceptions of the invasion and effects of *Senna obtusifolia*

According to local people, the rapid invasion of *Senna obtusifolia* into their environment is mainly promoted by livestock and human activities. These results are similar to the work of Gebreyesus (2017), who found that the proliferation of *Senna obtusifolia* is caused by the encroachment of extensive and illegal cultivation as well as grazing practices in various forms. This proves that these populations have a good perception of their environmental resources (Gaoue *et al.*, 2011).

In the Sahelian zone, according to the respondents, animals are the most important factors in the spread of *Senna obtusifolia*. They add that this is due to overgrazing, as the scarcity of fodder, especially in the dry season, forces livestock to ingest the dry pods of *Senna obtusifolia* and spread its seeds as they move. According to Thiombiano *et al* (2012), the spread of *Senna obtusifolia* is facilitated by cattle. In the Sudanian zone, respondents claim that man is the main contributor to the spread of *Senna obtusifolia*. This is due to the fact that many natural habitats have been occupied and transformed into fields, thus provoking the proliferation of *Senna obtusifolia*. According to Burgess *et al* (2004), more than 60-80% of natural habitats in West African countries have been transformed into agricultural and residential areas. These transformations have strongly contributed to the establishment and proliferation of invasive species (Pyšek *et al.,* 2010; Vila and Ibáñez, 2011). A similar finding was made by Archibald *et al.* (2020), who reported that humans are among the factors driving biological invasions. Furthermore, the majority of informants were unanimous that the invasion of *Senna obtusifolia* has led to the decline of several native and highly palatable species such as *Alysicarpus ovalifolius*, *Andropogon gayanus*, *Brachiaria lata*, *Pennisetum pedicellatum* and *Zornia glochidiata*. Our results corroborate those of Randell (1995); Solomon (2015); Gebreyesus (2017) who found that *Senna obtusifolia* is an aggressive pasture invader and can completely out-compete palatable grass species and eradicate the growth of pasture species. As a result, pasture invasion by this species reduces forage productivity, leading to forage shortages for livestock (Gebreyesus, 2017).

4.2. Local population adaptation strategies for forage acquisition in areas invaded by *Senna obtusifolia*

People have developed several strategies to save the little palatable forage available. Mowing herbaceous fodder, harvesting woody fodder, harvesting and conserving crop residues and growing fodder crops are the methods used by respondents to obtain fodder. In the Sudanian zone, the harvesting and conservation of crop residues remains the most dominant practice (Sanfo *et al.,* 2020; Zampaligré *et al.,* 2020), where crop residues play a major role in livestock feed due to the low productivity of rangelands and the increase in cultivated areas. These residues are mainly cereal straws and legume tops (Drabo *et al.,* 2001; Kiéma *et al.,* 2012). By contrast, in pastoral areas of the Sahel, mowing of herbaceous forage and crop residues are the most common practices. Our results are similar to those of Ouédraogo *et al.* (2021), who found that fodder resources exploited in the Sahel are mainly herbaceous fodder and crop residues left in the field after harvesting.

4.3. Effect of climatic zone on use of *Senna obtusifolia*

In this study, the use of *Senna obtusifolia* differed from one zone to another. In the Sudanian zone, the leaves are the most widely used organ of the species. They are used in the diet, as reported by Nacambo *et al.* (2021), who indicate that the leaves are eaten like vegetables. *Senna obtusifolia* leaves are highly valued by local populations (Abakar *et al.*, 2019), as they are rich in vitamins and proteins that can improve the nutritional quality of food (Dansi *et al.*, 2008; Nacambo *et al.*, 2021) and enable people to vary their diet and combat hunger during the lean season (Millogo, 2001; Batawila *et al.,* 2005).

The lack of knowledge about the use of other parts of *Senna obtusifolia* could be explained by the choice made by local people in the use of plants to satisfy their needs. This choice is based on the availability and diversity of plant resources in the area. The same observations were made by Cotton (1996) and Schmidt *et al.* (2005), who found that high floristic diversity leads to preferences in species use among populations. In the Sahelian zone, however, *Senna obtusifolia* leaves are used along with other parts of the species. These include the dry stems used to make seccos, hut roofs and granaries. They are also used as fertilizer in the fields or as firewood or skewers (Gannouka, 2021; Nacambo *et al.,* 2021). The roots and seeds of *Senna obtusifolia* are often used in food and medicine. This means that *Senna obtusifolia* is becoming increasingly "accepted" by populations. Indeed, it has been shown that when a species starts to become invasive in a region, firstly, the species can be perceived mainly negatively if it colonizes crops (Souley *et al.,* 2020). Secondly, people may gradually appropriate the plant and test different uses until they find a use, at which point the perception becomes less negative (Holmes *et al.*, 2009). The gradual use of different parts of *Senna obtusifolia* could be explained by the low availability of plant resources due to deteriorating environmental conditions and arid climate in this area. This scarcity of resources in the Sahelian zone is confirmed by Masson-Delmotte *et al.* (2018) and Ayantunde *et al.* (2020), who have shown that the Sahelian zone is characterized by high sensitivity to climate variability, vulnerability to drought and diminishing plant resources. This scarcity of resources leads populations to develop adaptation strategies such as the use of invasive species to replace those that have disappeared. More specifically, local populations have developed numerous traditional uses of *Senna obtusifolia* as coping strategies to the negative effects of its invasion into grazing areas (Ouedraogo *et al.*, 2021). A similar case was found with *Sida cordifolia L.*, an invasive species in Niger that has gradually replaced certain herbaceous plants, notably *Andropogon gayanus* Kunth and *Eragrostis tremula* Hochst.ex Steud. in the manufacture of seccos, hut roofs and granaries (Souley *et al.*, 2020).

4.4. Effects of socio-demographic groups on the use of *Senna obtusifolia*

The use of *Senna obtusifolia* differs from one socio-cultural group to another. In the Sudanian zone, the Bobo, Dogossin and Tieffo use *Senna obtusifolia* leaves for food (sauce). According to informants, the Bobo learned to use *Senna obtusifolia* leaves from their parents. They call *Senna obtusifolia* "kikir". This means a plant with a bitter taste. To prepare it, the Bobo takes care to boil the leaves and pour off this first water before using a second water for its preparation. This makes the sauce less bitter. Dogossin and Tieffo, on the other hand, learned to make sauce from *Senna obtusifolia* leaves with the Bobo. The use of this species is transmitted through cultural mixing (Batawila *et al.*, 2005; Adjatin, 2006). Hence the name of the sauce made from *Senna obtusifolia* leaves, attributed to the Bobo and known as "Bobo sauce". Sometimes the species is called tiguinfara, sometimes "kikir", a name borrowed from the Bobo. The use of the same name by two or more socio-cultural groups suggests that they occupy the same geographical area or have common historical links, as reported by Dansi *et al.* (2008). For example, Mossi and Peulh from both climatic zones use the same vernacular name for the species.

The Mossis and Peulhs of the Sudanian zone also use *Senna obtusifolia* leaves in their diet. The Peulh use it to prepare sauce. The Mossi use Senna obtusifolia leaves either to prepare sauce, or couscous made from a mixture of maize or millet with *Senna obtusifolia* leaves, or "youngou" made from boiled and pressed *Senna obtusifolia* leaves seasoned with oil and salt. However, in the Sahelian zone, the Bela, Mossi, Fulani, Gourmantché and Sonrai also use *Senna obtusifolia* leaves in their food (sauce, couscous, youngou). Knowledge of how to use *Senna obtusifolia* also varies widely from one socio-cultural group to another. The Bela use the leaves and roots to treat stomach upsets and urinary tract infections. This result is in line with previous studies by Doughari *et al* (2008), who found that *Senna obtusifolia* leaves are also used locally as a remedy for stomach aches, urinary tract infections, dysentery, diarrhea, fever and coughs. The Bela also use the dried stems to make seccos, hut roofs and granaries for their own use or for sale. Some Bela use the stems to make fire. Bela, Mossi, Foulani, Sonrai and Gourmantché learned this activity from their parents. Knowledge is passed down from generation to generation (Goudégnon *et al.,* 2018). Bela claims that seccos made from the dry stems of *Senna obtusifolia* are more resistant than those made from the stems of *Andropogon gayanus*. The Sonrai and Gourmancthé use the dried stems to make seccos, or hut roofs. The Gourmantchés also use the dried species as fertilizer in their fields. Mossi and Foulani use dried *Senna obtusifolia* stems to make seccos, huts and granaries. The Mossi also use the leaves and roots to treat stomach ailments, dysentery and diarrhea. This is in line with the study by Nacambo *et al* (2021), who reported that many parts of *Senna obtusifolia* are used in the treatment of certain diseases. In this respect, in addition to the leaves and stems, the Mossi also use the seeds as a substitute for coffee, in line with the findings of Doughari *et al.* (2008) who show that *Senna obtusifolia* seeds are often roasted and boiled in water to produce tea or coffee. These findings corroborate those of others in West Africa who have shown the influence of socio-cultural groups on plant uses (Ekué *et al.,* 2010; Koura *et al.,* 2011). Within socio-cultural groups, socio-demographic characteristics such as age and gender have no influence on the use of *Senna obtusifolia*. This could be due to the fact that uses are perceived at community level rather than at individual level.

5. Partial conclusion

This study provided information on the knowledge and use of *Senna obtusifolia* as an invasive species. Populations have reacted to the invasion of *Senna obtusifolia* as both victims and beneficiaries, and consider the acceptance of *Senna obtusifolia* as an integral part of the rural ecosystem to be an inevitable outcome. People's knowledge of the use of *Senna obtusifolia* varies greatly from one socio-cultural group to another. This confirms our first hypothesis, which states that endogenous knowledge of *Senna obtusifolia* is associated with a specific socio-cultural group. The use of *Senna obtusifolia* is widespread among socio-cultural groups. However, the Bela are more familiar with the use of *S. obtusifolia* than other socio-cultural groups. Our second hypothesis, that the use value of the species differs according to climatic zone, is confirmed. Almost all parts of this species are used for many purposes. People use *Senna obtusifolia* for food, traditional medicine and as a source of income. The species is more widely used in the Sahelian zone than in the Sudanian zone. Our third hypothesis, which stipulates that the forms of valorization of the species are more diversified in the Sudanian zone, is invalidated. With regard to the scarcity of plant resources, all informants know and use *Senna obtusifolia* to replace species that have disappeared. This finding could help stakeholders to guide policy formulation and better value invasive species such as *Senna obtusifolia* at community rather than local level.

CHAPTER III: DETERMINANTS OF THE PROLIFERATION OF *SENNA OBTUSIFOLIA* (L.) H.S IRWIN & BARNEBY AND IMPACT ON ECOSYSTEM SERVICES

1. Introduction

Importation, introduction, acclimatization, reproduction and dispersal are a set of processes leading to the establishment of a species and its colonization of new ecological environments (Yohann, 2015). These processes may be biological, environmental or intrinsic to the invasive species (El-Barougy *et al.*, 2020; Leal *et al.*, 2022). In the context of climate change, a previously non-invasive species may become invasive, and many native species shift their geographic distributions, entering areas where they were previously absent (Hellmann *et al.*, 2008). Through their proliferation, invasive plants induce significant changes in the composition, structure and/or functioning of ecosystems. They can compromise the ecological balance of habitats (Triplet, 2012). They can harm biodiversity and other ecosystem services by competing with and sometimes replacing native ecosystems (Nzigidahera, 2017). In addition to these ecological nuisances, economic and health impacts are frequently encountered (Triplet, 2012). In Burkina Faso, *S. obtusifolia* has invaded Sahelian and Sudanian pastures, leading to a considerable reduction in biodiversity and forage. Unfortunately, little attention has been paid to date to the causes of the species' proliferation and expansion, or to the effect of its invasion on ecosystem services. Specifically, it involves :

- Determine the role of climate, land use, topography and their interaction on biomass, density and height of *S. obtusifolia* ;
- Determine the influence of climate, land use and topography on *S. obtusifolia* yield parameters;
- Assessing the impact of *S. obtusifolia* invasion on the attributes of herbaceous vegetation in pastoral areas.
- Assessing the impact of *S. obtusifolia* invasion on ecosystem services

The assumptions supporting this activity state that:
- Climate is the most important factor in the invasion of *S. obtusifolia* ;
- Topography positively influences the yield parameters of *S. obtusifolia* ;
- Taxonomic richness decreases as the level of invasion by *S. obtusifolia* increases;
- Pastoral value declines as the level of invasion by *S. obtusifolia* increases.

2. Methodology

2.1. Sampling plan

A preliminary field survey was carried out to identify and select potential study sites. In order to understand the role of climate in the proliferation of the species, surveys were carried out in two climatic zones: the Sahelian and Sudanian zones. Within each climatic zone, land use types (pastoral/ fallow areas) were identified (Table 5, Appendix 2).

The distinction between pastoral and fallow areas was intended to reflect the effect of land use on the expansion of *S. obtusifolia*.

Fallow land is an area of crops temporarily set aside for the natural restoration of soil fertility. It can be used for grazing or as a range for animals (Akpo *et al.*, 2002; law n° 034-2002/an). In the extensive livestock farming system, fallow land constitutes almost all of the natural fodder preferred by livestock farmers during the wet season (Koutou *et al.*, 2016).

According to law n°034-2002/an, which sets out the orientation law for pastoralism in Burkina Faso, pastoral zones are areas set aside and open to animal grazing.

The stratified, presence-oriented inventory method was adopted (Yameogo *et al.,* 2013; Glèlè Kakaï *et al.*, 2016), taking into account four (4) invasion levels of *S. obtusifolia*. These invasion levels are defined according to those of Gebrekiros and Tessema 2018:

Invasion stage 1 (control level): corresponds to the pre-contamination stage, during which no *S. obtusifolia* plants are present in the environment.

Invasion stage 2 (low level or A): corresponds to the stage from contamination/introduction to the beginning of establishment, but the invasive species always remains subordinate to other species in the environment; during this stage, the invasive plant has a cover varying from 1 to 30%.

Invasion stage 3 (medium or B level): corresponds to the stage where the invasive plant occupies an increasingly higher proportion of **the ground** than other species, due to its intrinsic capacity (rapid reproduction) and the selection of other herbaceous species eaten by livestock; during this stage, the invasive plant has a cover varying from 30 to 60%.

Stage 4 of invasion (high level or C): this is reached when the invasive plant has become abundant and dominant, even monospecific, or by presenting a cover rate of over 60%, for which *S. obtusifolia* becomes an ecological threat (Richardson *et al.* (2000).

A total of 160 plots of 100 m each were set up^2. That's 80 plots per study site, including 40 in pastoral zones and 40 in fallow land, with 20 plots per topographical unit.

In order to determine the role of topography on the invasion of the species, surveys were carried out in glacis and lowlands. According to Kadéba *et al.* (2015), *S. obtusifolia* is one of the characteristic species of lowland plant communities in the Sahelian zone of Burkina Faso. In each of the topographic levels (glacis and lowland), four (04) 100 m plots^2 (Ouédraogo, 2004; Zerbo, 2011), including one control and each of the other three corresponding to a precise level of *S. obtusifolia* invasion were made using a stretched metric tape. Within each plot, three sub-plots were placed, according to the level of invasion defined for the plot in question. Each sub-plot is marked by a 1 m iron frame^2. The 1 m x 1 m size of the plots and the square shape help to avoid edge effects and facilitate the rapid establishment of plots in tall formations (Fournier, 1991, Porembsky, 2011). In the field, at the highest levels of invasion, *S. obtusifolia* is around 1m to 2m tall. The GPS coordinates of the plot were recorded.

Table 5Sampling plan

Climate zone	Sahelian/Sudanese		
Sites	Sambonaye Sideradougou		
Number of plots	80 80		
Type of land use	Pastoral/ fallow area Pastoral/ fallow area		
Number of plots	40 40 40 40		
Topographic level	Glacis/Bas-fond Glacis /Bas-fond		
Number of plots	20 20 20 20		
Invasion level	T A B C T A B C		
Number of plots	10 10 10 10 10 10 10 10		
Number of subplots	30 30 30 30 30 30 30 30		

2.2. Floristic surveys

Floristic data were collected from August to September in the Sahel zone and from October to November in the Sudanian zone. These periods (August to November) correspond to the leafing and flowering of several West African savannah grasses

(Fournier, 1983), including *S. obtusifolia*. Specific height and cover rate, average height and cover rate of the herbaceous stratum were recorded (Appendix 3) in plots 100 m² (10m x10m) in size (Ouédraogo, 2006; Zerbo, 2011).

An exhaustive list of herbaceous species in the plot was first drawn up, and each species assigned its abundance-dominance coefficient according to the Braun-Blanquet (1932) scale in the format modified by Williams (1989) and used by Thiombiano (2005).

The abundance-dominance coefficient assigned to the species is :

    5: coverage between 75 and 100% of the surface area

    4: overlap between 50 and 75% of the surface area

    3: overlap between 25 and 50% of the surface area

    2b: overlap between 15 and 25% of the surface area

    2a: overlap between 5 and 15% of the surface area

    1b: overlap between 3 and 5% of the surface area

    1a: overlap between 1 and 3% of the surface area

    1: 1 and 5% of the survey area

    +: a few individuals with less than 1% of the survey area.

    r: rare species, coverage less than 1% of survey area

A sample of each species was taken for identification and comparison with herbarium specimens from Joseph KI-ZERBO University. A code was assigned to each species whose scientific name was unknown. The scientific names of the species were verified and determined with the help of flora and botanical works (Berhaut, 1979; Poilecot, 1995; 1999; Bonnet *et al.*, 2005; Thiombiano *et al.*, 2012; Arbonnier, 2019;), and with the assistance of PhD students, technicians and teacher-researchers from the Laboratoire de Biologie et Écologie Végétales at the Université Joseph KI-ZERBO.

Within each plot, three (03) sub-plots of 1 m² (1 m x 1 m) were then randomly installed and the average height and total number of individuals (density) of *S. obtusifolia* were recorded (Appendix 2). In each sub-plot, biomass cutting operations were carried out where all grass was mowed flush to the ground (Grouzis, 1984; Alhassane *et al.*, 2018; Moussa *et al.*, 2018) using a sickle and placed on plastic sheeting. By sorting, *S. obtusifolia* was separated from the other grasses and the fresh weight of each pile was determined using a 50 kg ± 10 g precision electronic balance. Using a 5 kg ± 1 g precision electronic balance, a 100 g sample was taken from each pile. To obtain a constant dry weight, these samples were first dried in the field and then in the laboratory in an oven for 72 hours at 75°C, to avoid denaturing the sample's constituent elements, and then used to determine the dry biomass (Ouédraogo, 2009; Zerbo, 2011). To determine the functional traits of *S. obtusifolia*, five (5) individuals were randomly selected from each plot. For each of the five individuals, the height was measured, the number of pods counted, the number of seeds per pod counted, and the average seed weight per pod determined.

2.3. Analysis of floristic data

2.3.1. Biological types and phytogeographical

Biological and phytogeographical types have been used to characterize the herbaceous flora of pastoral zones. Biological types reflect variation in environmental conditions (Schmidt *et al.,* 2005). An abundance of therophytes in the study area could indicate an anthropogenic disturbance such as overgrazing (Sinsin, 2001). As for phytogeographical types, they highlight the distribution range of plant species in different environments (White, 1986).

Based on the entire floristic list of the sites, we have established the percentage of species belonging to each of the life forms (or biological types) of the plant described by Raunkier's (1934) classification in :
- Phanerophytes (Ph): plants whose shoots or persistent buds are located on the aerial axes more than 40 cm above the ground.
- Chaméphytes (Ch): plants with buds or persistent shoot tips close to the ground, on creeping or upright branches.
- Geophytes (G): plants whose persistent shoots or buds are sheltered in the ground during the off-season.
- Hemicryptophytes (H): plants whose replacement shoots or buds are located at ground level.
- Therophytes (Th): annual plants without persistent vegetative organs, propagated from year to year by seeds.

The nomenclature used for the phytogeographical types selected here follows that of the vascular plant catalog by Thiombiano *et al.* (2012). These are :

1. Species with a wide geographical distribution :

- Cosmopolites (Cos): species distributed in both tropical and temperate regions;
- Pantropical (Pan): species found in all tropical regions: Africa, America and Asia;
- Paleotropics (Pal): species found in tropical Africa, Asia, Madagascar and Australia;

2. African multi-regional species

- Afro-Malagasy (AM): species distributed in Africa and Madagascar;
- Afro-tropical (AT): species distributed in tropical Africa ;
- Sudano-Zambézian (SZ): species distributed in both African, Sudanian and Zambézian (SZ) environments.

This information was used to determine the various vegetation indices of the invaded habitats, to analyze the effect of the invasion on the pastoral value of the study sites through the establishment of a forage spectrum. The heaps of other herbaceous plants formed were used to establish their above-ground biomass. The pile of *S. obtusifolia* was used to establish the above-ground biomass *of S. obtusifolia.* APG IV (2016) nomenclature was used for systematic characterization of species into families.

2.3.2. Pastoral diversity and values

PCord 6 software was used to calculate alpha diversity indices such as species richness, Shannon diversity, Pielou equitability and Simpson diversity. Specific richness (SR) with $RS=$ number of species per plot, indicates the number of species encountered in an environment. Shannon's diversity index (H), with :

$$H = - \Sigma Pi * \ln(Pi) ;$$

Where Pi is the relative abundance of the ith species in the plot. This index quantifies the heterogeneity of an environment's species diversity. H generally varies on average from 0 to 5 bits. Its high values reflect species abundance, and therefore favorable environmental conditions. H=0 if all individuals in the group belong to the same species (Maguran, 2004).

Piélou's equitability (E) represents the equi-repartition of all individuals present by species.

$$E = H/\ln(S)$$

It varies from 0 to 1 and indicates a regular distribution of individuals between species with values close to 1, and the dominance of one species or a small number of species with values close to 0.

Simpson's diversity index (D) with :

$$D = \Sigma(Pi^2),$$

Where pi is the relative abundance of the ith species in the plot. Simpson's index measures the heterogeneity of species abundance. D=0 if all individuals in the stand belong to the same species. The higher the value, the greater the diversity (Maguran, 2004).

β-diversity was assessed using Sørensen's similarity index (Cs) based on species abundance and presence/absence. This made it possible to compare the similarity between the two types of land use (fallow and pasture).

$$Cs = \frac{2J}{2J+a+b}$$

Where, j = number of common species, a = number of species found only in site A and b = number of species found in site B. This index varies between 0 and 1: a value close to 1 indicates high similarity between types and hence low beta-diversity (Maguran 2004).

To determine the pastoral value of the study sites, we first compiled a floristic list of the species identified by invasion level in the ten (10) pastoral zone sites. On the basis of the literature, each species was then assigned either a specific quality index, for herbaceous species, or a palatability qualifier (NA: Not palatable; PA: Little palatable; A: Palatable and TA: Very palatable) in the case of juvenile woody species (Kiéma, 2002; 2007; 2008; Sawadogo, 2011). The specific quality index (Is) translating the bromatological value of each herbaceous species (Akpo *et al.*, 2002) was determined. It is based on a scale of four classes ranging from 0 to 3 (Akpo *et al.*, 2000; Akpo *et al.*, 2002; Sawadogo, 2011; Ngom *et al.*, 2012): 0: Species of No Pastoral Value (SVP), 1: Species of Low Pastoral Value (FVP), 2: Species of Medium Pastoral Value (MVP), 3: Species of Good Pastoral Value (BVP).

With regard to the woody regeneration contained in the surveys, unpalatable species are classified as SVP, poorly palatable species as FVP, palatable species as MVP and highly palatable species as BVP. Bechir *et al* (2015) used a scale of six classes, from 0 to 5, with 0: no pastoral value; 1: poor pastoral value; 2: average pastoral value; 3: fairly good pastoral value; 4: good pastoral value and 5: very good pastoral value. By comparing the four-class scale with the six-class scale, classes 2 and 3 on the one hand, and classes 4 and 5 on the six-class scale on the other, were assigned to classes 2 and 3 respectively on the four-class scale. The qualifiers NA, PA, A and TA were assigned to classes SVP, FVP, MVP and BVP respectively. These different species groups (BVP, MVP, FVP and SVP) or index classes constitute the forage species categories. A literature review was used to assign specific quality indices and palatability qualifiers to the species (Akpo *et al.*, 2000; Kiema, 2002; Sare, 2004; Kiema, 2007; Kiema, 2008; Sawadogo, 2011; Ngom *et al.*, 2012). Finally, the average cover rate of each species was determined in the plots. The rate of cover of all species of the same forage or non-fodder species category as a function of invasion levels was designed. This made it possible to represent a forage spectrum, indicative of the cover rate of each forage value class, as a function of invasion level.

To identify the most dominant herbaceous species, the relative importance of each species was calculated according to the level of invasion in each study area. It was calculated from the sum of relative frequency and relative dominance, which were obtained using the following formulas:

$$\text{Frequency} = \frac{\text{Nombre de placeaux dans lequel l'espèce apparait}}{\text{Nombre total de placeaux}} \; x100$$

$$\text{Relative frequency} = \frac{\text{Frequence d'une espèce}}{\text{Somme de toutes les frequences}} \; x100$$

$$\text{Relative dominance} = \frac{\text{Total des taux de recouvrement d'une espèce}}{\text{Total des taux de recouvrement de toutes les espèces}} \; x\,100$$

Relative Importance = Relative Frequency + Relative Dominance

Dominant herbaceous species are those with a relative importance value greater than 5 (IR ≥ 5).

The dry biomass and average density of *S. obtusifolia* were then calculated. The aerial biomass of the herbaceous stratum corresponds to the weight of dry matter of all aerial organs per unit area. It is expressed in tonnes of dry matter per hectare (Zerbo, 2011). The Shapiro-Wilk normality test revealed that the data were not normally distributed. Consequently, the non-parametric Mann Whitney test at the 5% threshold was used to test the influence of climatic zone, land use pattern and topography on *S. obtusifolia* invasion.

To assess the impact of climate, land-use types, topography and their interaction (explanatory variables) on the biomass, height and density of *S. obtusifolia* (response variable), Generalized Linear Model (GLM) analyses with poisson and Gamma (biomass) errors were carried out.

A multiple comparison of the biomass *of S. obtusifolia,* the biomass of other species in the herbaceous stratum and species richness as well as diversity indices (Shannon diversity and Piélou equitability) between invasion levels was carried out using Tukey's test at the 5% significance level. Pearson's correlation test was used to test the correlation between biomass and height of *S. obtusifolia.* The Wilcoxon test was used to test variation between *S. obtusifolia* functional traits in the two climatic zones. Spearman's Correlation Matrix was used to test correlation between *S. obtusifolia* functional traits. Histograms were used to represent biological and phytogeographical types using raw (number) and weighted (overlap) spectra. All these analyses were carried out using R software version 4.1.0.

3. Results

3.1. Effects of climatic zone, topography, type of land use and their interactions on density and biomass *of Senna obtusifolia*

Overall, climatic zone, land use type and topography and their interaction had a significant effect ($p < 0.05$) on *S. obtusifolia* density and biomass (Table 6). Indeed, Pearson's correlation test reveals that there is a positive correlation between height and biomass of *S. obtusifolia* ($r=60.35\%$).

Table 6Effects of climatic zone, topography, type of land use and their interactions on density and biomass of *Senna obtusifolia*

Environmental factors	Density of *S. obtusifolia*		Biomass of *S. obtusifolia*	
	Z value	Pvalue	Zvalue	Pvalue

Climate	-20,443	0,001	3,053	0,002
TUT	-11,442	0,001	2,333	0,019
Topo	-6,561	0,012	1,475	0,014
Climate *TUT	7,667	0,001	-1,786	0,05
Climate *Topo	8,597	0,023	-1,544	0,012
TUT*Topo	3,872	0,04	-1,883	0,03
Climate*TUT*Topo	-4,174	0,001	1,590	0,01

TUT: Type of land use; Topo: topography. Significance at 0.05 level

3.2 Effects of land use on *Senna obtusifolia* density and biomass in different climatic zones

In the Sahel, *S. obtusifolia* has a high density, but the individuals have a low biomass. In contrast, the Sudanian zone is characterized by a low density of *S. obtusifolia* individuals with a higher average biomass (Table7). The biomass of other grasses is higher in the Sudanian zone. Land use type has a highly significant effect (p < 0.05) on *S. obtusifolia* density and biomass. Pastoral areas are characterized by high density but low biomass of *Senna obtusifolia*, while fallow land is characterized by low density and high biomass of *S. obtusifolia*. The biomass of other herbaceous species is lower than that of *S. obtusifolia*. It is higher in the Sudanian than in the Sahel (Table 8).

In the Sudanian climatic zone, land use type has a significant effect (p<0.05) on *S. obtusifolia* density. However, it has no significant effect (p > 0.05) on biomass (Table 8).

Table 7Effect of climate on density and biomass of *Senna obtusifolia* and other grasses by climatic zone

	Sahelian	Sudanese		
Parameters	**M±E**	**M±E**	**W**	**P**
Density(ind/ha) of *S. obtusifolia*	91,47 ± 74,06.10⁴	13,90 ± 21,10.10⁴	295,23	<0,001
Biomass(t/ha) of S. *obtusifolia*	2,14 ± 2,09	3,04 ± 2,92	130,25	<0,001
Biomass (t/ha) of other grasses	1,45 ± 1,25	1,82 ± 1,75	12397	<0,001

Table 8Effect of land use type on density and biomass of *Senna obtusifolia* and other grasses by climatic zone

Zones		Pastoral zone	Fallow land		
		M±E	M±E	W	P
Sahel	Density(ind/ha) of *S. obtusifolia*	120,91 ±80,31.10⁴	62,04 ±53,21.10⁴	0,9034 3	<0,00 1
	Biomass(t/ha) of S. *obtusifolia*	1,73 ± 1,66	2,55 ± 2,40	2243,5	0,008
	Biomass (t/ha) of other grasses	1,26 ±1,24	1,71 ±1,56	4613	0,05
Sudanese	Density(ind/ha) of *S. obtusifolia*	20,23 ± 28,01.10⁴	7,57 ± 5,48.10⁴	2773,5	<0,00 1
	Biomass(t/ha) of S. *obtusifolia*	2,77 ± 2,52	3,30 ± 3,27	4201	0 ,034
	Biomass (t/ha) of other grasses	1,62 ±1,02	2,14 ±1,46	5082,5	<0,00 1

Zp: pastoral zone; Ind./ha: individuals per hectare; t/ha: tonne per hectare. * indicates a significant difference

3.3. Effects of topography on density and biomass of *Senna obtusifolia* in different climatic zones

Generally speaking, topography has a significant effect (p< 0.05) on the density and biomass of *S. obtusifolia* (Table 9). In both the Sahel and Sudanian zones, topography influences the density and biomass (p < 0.05) of *S. obtusifolia* (Table 9).

Table 9Effect of topography on density and biomass of *Senna obtusifolia* and other grasses according to climatic zones

Zones	Parameters	Lowlands	Glazing		
		M±E	M±E	W	P
Sahel	Density(ind/ha) of *S. obtusifolia*	11,03 ± 82,04.10⁵	7,264 ± 59,88.10⁴	5072, 5	<0,00 3
	Biomass(t/ha) of *S. obtusifolia*	2,34 ± 2,23	1,83 ± 1,51	4072	0 ,045
	Biomass (t/ha) of other grasses	1,67 ± 1,4	1,12 ±1,14	5158	<0,00 1
Sudanese	Density(ind/ha) of *S. obtusifolia*	19,14 ± 28,32.10⁴	8,65±61,68.10³	4562	<0,00 1
	Biomass(t/ha) of *S. obtusifolia*	3,45 ± 3,43	2,62±2,25	4251	0 ,049
	Biomass (t/ha) of other grasses	1,97 ± 1,33	1,76 ±1,18	4731	0,026

Ind./ha: Individuals per hectare; t/ha: tonne per hectare; * indicates a significant difference.

3.4.Effects of climatic zone on height and yield parameters of *Senna obtusifolia*

The height and yield parameters of *S. obtusifolia* vary significantly ($p < 0.05$) from one zone to another. Individual height, number of pods, average pod weight, number of seeds per pod and seed weight are higher in the Sudanian zone than in the Sahelian zone (Table 10). However, pod length does not vary from one zone to another.

The correlation matrix shows a medium correlation between the height of *Senna obtusifolia* and the number of pods (Table 11). The correlation is very strong between average pod weight and seed weight.

Table 10Height and yield parameters of *Senna obtusifolia* according to climatic zones

		H (cm)	NG/ind	LG (cm)	PmG (g)	Ng/G	Pg (g)
Sahel	M±sd	82,5 ± 35,76	6 ± 13,51	18 ± 1,98	1 ± 1,04	25 ± 4,00	0,018 ± 1,01
Sudanese	M±sd	121,0 ± 36,04	22 ± 52,78	18 ± 2,13	2 ± 0,79	25 ± 4,16	0,032 ± 0,6
Wilcoxon test		W=0,057 P <0,0001*	W=124 P <0,0001*	W=415 P =0,599	W=295 P<0,001*	W=282 P =0,1	W=319 P <0,0001*

Legend: H: height of each *S. obtusifolia* individual; NG: number of pods per individual; LG: pod length; PmG: average pod weight per individual; Ng/G: number; *Indicates a significant difference; M±sd: mean ± standard deviation.

Table 11Spearman's correlation matrix between functional traits of *Senna obtusifolia*

	H (cm)	NG	LG (cm)	PmG (g)	Ng /G	Pg (g)
H (cm)	1					
NG	**0,54**	1				
LG (cm)	0,21	0,26	1			
Pm G (g)	0,24	0,40	0,31	1		
Ng /G	0,26	0,35	0,36	0,25	1	
Pg (g)	0,18	0,33	0,18	**0,92**	0,02	1

Legend: H: height of each *S. obtusifolia* individual; NG: number of pods per individual; LG: pod length; PmG: average pod weight per individual; Ng/G: number of seeds per pod; Pg: seed weight. Values in bold represent significant correlations.

3.5. Effects of land use and topography on height and yield parameters of *S. obtusifolia*

In the Sahelian zone, the height and yield parameters of *S. obtusifolia* vary significantly (p< 0.05) across both land-use types and topographic levels. Average pod weight (PmG) and seed weight (Pg) were higher in the fallow than in the pastoral zone. As for topography, the height of *S. obtusifolia* and the number of pods per individual measured in the lowlands are higher than those found on the glacis (Table 12).

Table 12Effects of land use and topography on height and yield parameters of *Senna obtusifolia* in the Sahelian zone

	H(cm)	NG/ind	LG (cm)	PmG(g)	Ng/G	Pg (g)
Zp	80 ± 32,79	5 ± 8,15	18 ± 2,28	0,5± 0,89	25 ± 2,85	0,2 ± 0,57
Fallow land	88 ± 37,80	6 ± 16,53	18 ± 1,71	2 ± 1,03	25,5± 4,72	1± 1,19
Wilcoxon test	W=10118	W=9779,5	W=8859	W=13086	W=10456	W=13376
	P=0,08	P=0,02*	P=0,82	P<0,0001*	P=0,022*	P<0,0001*
Basfond	92,56±34,09	6 ± 15,51	18,0± 1,99	1 ± 1,03	25 ± 4,11	0,3± 0,73
Glazing	85,69±37,17	5 ± 7,69	18,5 ±1,96	1 ± 1,03	25± 3,09	0,5± 1,22
Wilcoxon test	W=10676	W=7262	W=8037	W=7147	W=8994	W=7465
	P=0,014*	P=0,004*	P=0,09	P=0,2	P=0,85	P=0,009*

Legend: Zp: pastoral zone; H: height of each *Senna obtusifolia* individual; NG: number of pods per individual; LG: pod length; PmG: average pod weight per individual; Ng/G: number of seeds per pod; Pg: seed weight.*indicates a significant difference.

In the Sudanian zone, the number of pods and average pod weight vary significantly (p< 0.05) with land use type. In fallow land, the number of pods per individual and average pod weight are higher than in pastoral land. In the lowlands, the number of pods is significantly (p< 0.05) higher than in the glacis (Table 13).

Table 13Effects of land use and topography on height and yield parameters of *Senna obtusifolia* in the Sudanian zone

	H (cm)	NG/ind	LG (cm)	PmG (g)	Ng/ G	Pg (g)
Grazing	118,5±42,65	17 ± 8,15	18 ± 2,28	1,9 ± 1,03	27± 2,85	0,5 ± 0,57
Fallow land	128,5±27,03	25,5±16,57	18 ± 1,71	2 ± 0,89	27 ± 4,72	1± 1,19

Wilcoxon test	W=10248	W=14712	W=12350	W=13520	W=10951	W=13848
	P=0,018*	P<0,0001*	P=0,14	P=0,1	P=0,69	P<0,0001*
Basfond	123,5 ± 39,9	29 ± 69,9	18 ± 2,15	2 ± 0,83	28 ± 3,9	1 ± 0,67
Glazing	116 ± 30,6	19 ± 17,6	18 ± 2,08	2 ± 0,75	27 ± 4,3	1 ± 0,59
Wilcoxon test	W=9378	W=7432	W=10048	W=11702	W=8621	W=11384
	P=0,012*	P<0,0001*	P=0,10	P=0,52	P <0,12	P=0,85

Legend: H: height of each *Senna obtusifolia* individual; NG: number of pods per individual; LG: pod length; PmG: average pod weight per individual; Ng/G: number of seeds per pod; Pg: seed weight. *indicates a significant difference

3.6. Taxonomic richness (family, genus, species) of herbaceous vegetation

The flora inventoried at both sites (Sahelian and Sudanian zones) comprises 241 species collected in 160 plots, including 185 herbaceous species and 56 juvenile woody species (Appendix 3). Herbaceous species belong to 163 genera and 54 families. The most dominant families (Table 14) are Poaceae (16.34%), Fabaceae (13.42%), Cyperaceae (7.45%) and Rubiaceae (7.40%).

Table 14 Taxonomic richness of herbaceous vegetation in the two climatic zones

Families (%)	Sahel	Sudanese	Families (%)	Sahel	Sudanese
Arecaceae	-	0,04	**Euphorbiaceae**	0,42	1,35
Acanthaceae	0,77	0,98	**Fabaceae**	19,29	17,37
Amaranthaceae	2,82	5,06	**Lamiaceae**	0,28	6,74
Anacardiaceae	0,14	0,13	**Loganiaceae**	-	0,89
Annonaceae	-	0,09	**Malvaceae**	8,74	14,95
Antheriaceae	0,07	0,49	**Marsileaceae**	0,14	-
Apocynaceae	-	0,49	**Meliaceae**	-	0,18
Araceae	0,28	-	**Molluginaceae**	0,07	0,04
Arecaceae	0,28	-	**Moraceae**	-	0,18
Asclepiadaceae	0,85	0,09	**Nyctagynaceae**	0,07	0,44
Asteraceae	3,1	5,94	**Onagraceae**	0,63	0,49

Balanitaceae	0,56	-	Oxalidaceae	0,07	1,6
Boraginaceae	-	0,09	Phyllanthaceae	-	1,2
Capparidaceae	0,07	0,13	Poaceae	40,15	16,34
Caryophyllaceae	0,35	0,22	Polygalaceae	0,14	0,44
Celastraceae	-	0,04	Portulacaceae	0,07	0,04
Chrysobalanaceae	-	0,04	Rhamnaceae	0,28	0,09
Cochlospermaceae	-	0,05	Rubiaceae	7,61	7,36
Combretaceae	0,42	1,73	Rutaceae	-	0,31
Commelinaceae	0,45	2,71	Sapotaceae	-	0,44
Convolvulaceae	2,89	2,22	Scrophulariaceae	0,63	0,09
Cucurbitaceae	0,14	-	Solanaceae	0,07	0,31
Cyperaceae	8,1	7,23	Verbenaceae	-	0,04
Dioscoreaceae	-	0,22	Zygophyllaceae	0,07	0,09
Ebenaceae	-	0,23			

3.7. Biological and phytogeographical types of inventoried species

The raw and weighted biological spectra highlight the representativeness and coverage of the herbaceous species sampled in the two climatic zones (Figure 10 and 11). In the Sahelian zone, the biological spectra of herbaceous plants at the study sites show a clear predominance of therophytes, with an average of 80% of species compared with other biological types (Figure 10). Therophytes are followed by phanerophytes, then champephytes and hemicryptophytes (Figure 10). The same trend is observed in the Sudanian zone, with therophytes accounting for 80% of species compared with other types (Figure 11).

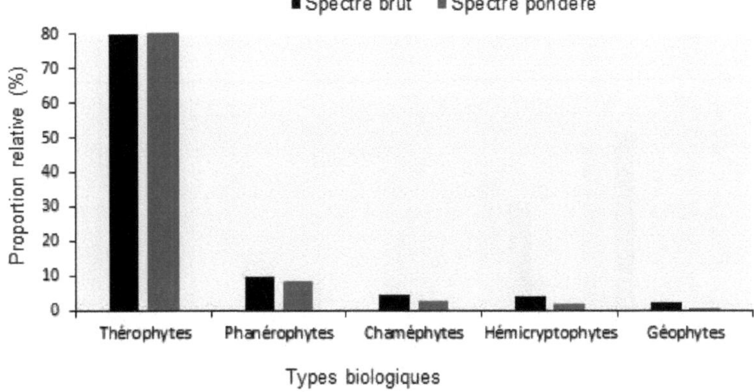

Figure 10Spectrum of biological types in the Sahelian zone

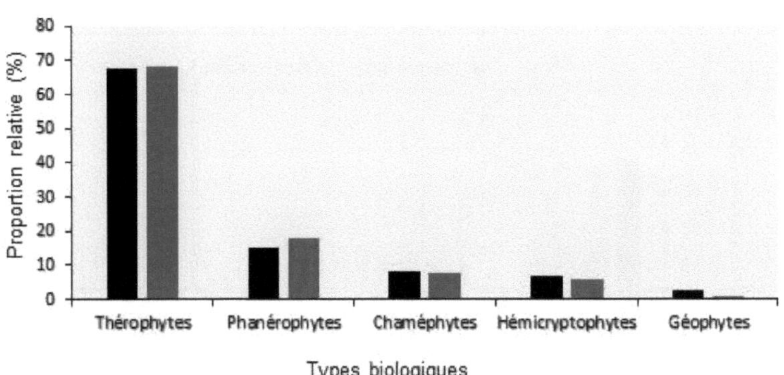

Figure 11Spectrum of biological types in the Sudanian zone

Depending on their phytogeographical affinities, plants recorded in the Sahelian zone are mainly of pantropical (Pan) origin, with an average of 50% of species, Afro-tropical (AT) around 35% of species, Afro-Malagasy and Paleotropical around 6% of species (Figure 12). In the Sudanian zone, pantropical and Afro-tropical species are the most represented, with average percentages of 65% and 25% respectively (Figure 13).

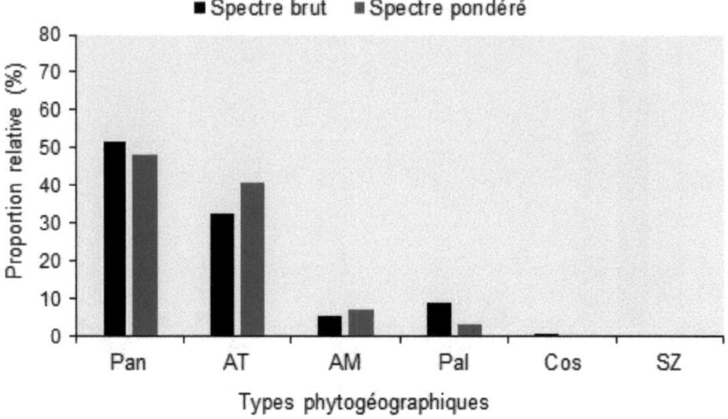

Figure 12Spectrum of phytogeographical types in the Sahelian zone

AT: Afro-tropical; AM: Afro-Malagasy; SZ: Sudano-Zambézian; COS: Cosmopolitan; Pal: Paleotropical; Pan: Pantropical

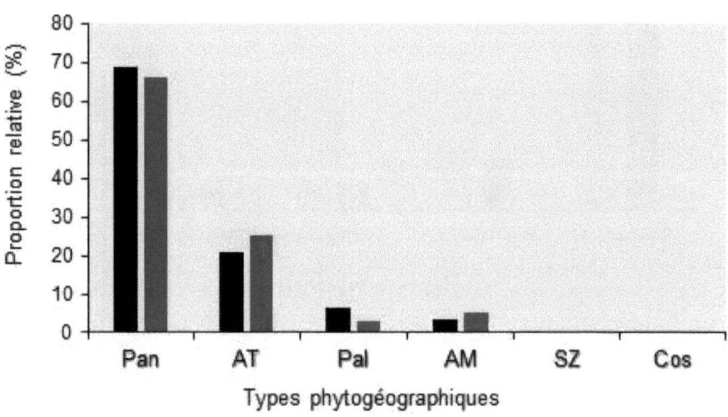

Figure 13Spectrum of phytogeographical types in the Sudanian zone

AT: Afro-tropical; AM: Afro-Malagasy; SZ: Sudano-Zambézian; COS: Cosmopolitan; Pal: Paleotropical; Pan: Pantropical

3.8. Diversity and similarity of invasion levels according to climatic zones, land use types and topography

3.8.1. Diversity in invasion levels according to land use type

In the Sahelian zone, and more specifically in pastoral areas as well as in fallow land, analyses show that the invasion of *Senna obtusifolia* has a significant effect on the mean number of species (ddl=3; F=13.01; P < 0.0001), diversity (ddl. =3; F= 34.24; P < 0.0001), and the distribution of these species (ddl. = 3; F= 30.4; P < 0.0001) as a function of invasion levels (Table 15). Equitability decreases significantly (ddl =3; F=21.34; P < 0.0001) from the control level to the highest invasion level (C). The number of species is higher in the control levels and decreases as the invasion level increases. The same trend can be observed in the Sudanian zone.

Table 15Distribution of floristic diversity according to level of invasion in land use type by climatic zone

Climate zones	TUT	Invasion level	S	H	D	E
			M±E	M±E	M±E	M±E
		T	23,6 ± 8,98[a]	2,30 ± 0,38[a]	0,86 ± 0,05[a]	0,74± 0,06[a]
	ZP	A	18,0 ± 6,16[b]	1,94 ± 0,35[a]	0,80 ± 0,07[a]	0,68 ± 0,08[a]
		B	12,9 ± 5,02[c]	1,40 ± 0,55[b]	0,6 ± 0,24[b]	0,55± 0,19[b]
		C	6,0 ± 2,54[d]	0,6 ± 0,36[c]	0,27 ± 0,17[c]	0,33 ± 0,14[c]
Sahelian		T	31,8 ± 12,16[a]	2,37 ± 0,43[a]	0,83± 0,11[a]	0,70 ± 0,08[a]
	Fallow land	A	22,0±4,16[b]	2,04 ± 0,24[a]	0,80 ± 0,04[a]	0,66± 0,04[a]
		B	16,1± 4,25[b]	1,62 ± 0,31[b]	0,68 ± 0,1[b]	0,59± 0,09[a]
		C	6,5 ± 2,80[c]	0,57 ± 0,27[c]	0,27 ± 0,15[c]	0,3 ± 0,15[b]
		Pvalue	< 0,0001	< 0,0001	< 0,0001	< 0,0001
		T	44,3±12,14[a]	2,85 ± 0,36[a]	0,90± 0,05[a]	0,76 ± 0,07[a]
	ZP	A	28,5 ± 8,32[b]	2,4± 0,2[b]	0,86 ± 0,025[a]	0,73 ± 0,04[a]
		B	21,5 ± 5,19[b]	2,1 ± 0,23[b]	0,8 ±0,05[a]	0,69 ±0,43[a]

Sudanese		C	$8,6 \pm 3,1^c$	$1,13\pm 0,38^c$	$0,5 \pm 0,18^b$	$0,53 \pm 0,15^b$
		T	$37,5 \pm 10,90^a$	$2,84\pm 0,30^a$	$0,90 \pm 0,04^a$	$0,79 \pm 0,06^a$
	Fallow land	A	$29,3 \pm 7,08^b$	$2,6 \pm 0,16^a$	$0,89 \pm 0,02^a$	$0,78 \pm 0,05^a$
		B	$23,0 \pm 5,03^c$	$2,42 \pm 0,22^b$	$0,8 \pm 0,04^a$	$0,75 \pm 0,07^a$
		C	$13,1 \pm 5,25^d$	$1,37 \pm 0,38^c$	$0,58 \pm 0,16^b$	$0,55 \pm 0,11^b$
	P-value		< 0,0001	< 0,0001	< 0,0001	< 0,0001

Comparisons are made by column (diversity indices for each type of land use are compared according to invasion level). Superscript letters indicate significant differences. Values with the same letters indicate no significant difference.
ZP: Pastoral zone
S: species richness; H: Shannon diversity index; D: Simpson diversity index; E: equitability Piélou, M±E: mean plus or minus standard deviation

3.8.2. Diversity as a function of invasion levels and topography

In both the Sahelian and Sudanian zones, the number of species decreases from the control level to the most invaded level. The same trends are observed in terms of species diversity and distribution (Table 16).

Table 16Distribution of floristic diversity by level of invasion and topography, by climatic zone

Climate zones	Topo	Invasion level	S	H	D	E
			M±E	M±E	M±E	M±E
		T	$20,8\pm 6,03^a$	$2,15 \pm 0,32^a$	$0,83\pm 0,07^a$	$0,72\pm 0,08^a$
	Lowlands	A	$18,8 \pm 5,02^a$	$1,95 \pm 0,27^a$	$0,79\pm 0,06^a$	$0,67\pm 0,07^a$
		B	$12,5\pm 4,22^b$	$1,44 \pm 0,38^b$	$0,63\pm 0,17^b$	$0,58\pm 0,14^b$
		C	$5,5 \pm 2,91^c$	$0,57\pm 0,40^c$	$0,27\pm 0,19^c$	$0,33\pm 0,18^c$
Sahelian		T	$38,2 \pm 8,78^a$	$2,71\pm 0,16^a$	$0,90\pm 0,23^a$	$0,75\pm 0,05^a$
	Glazing	A	$22,8 \pm 6,65^b$	$2,10\pm 0,35^a$	$0,81\pm 0,06^a$	$0,68\pm 0,07^a$
		B	$17,1\pm 4,36^b$	$1,61 \pm 0,51^a$	$0,66\pm 0,20^b$	$0,57\pm 0,16^b$
		C	$6,9 \pm 2,18^c$	$0,60 \pm 0,21^a$	$0,26\pm 0,12^c$	$0,32\pm 0,09^c$
	Pvalue					

			S (M±E)	H (M±E)	D (M±E)	E (M±E)
		T	37,8±8,4 3[a]	2,73± 0,23[a]	0,90±0,0 4[a]	0,76±0,0 7[a]
	Lowlands	A	30,7± 8,48[a]	2,60± 0,14[a]	0,88±0,0 2[a]	0,77±0,0 5[a]
		B	21,3± 4,87[b]	2,20 ±0,25[b]	0,82±0,0 5[a]	0,73±0,0 7[a]
		C	11,0± 4,11[c]	1,25 ± 0,37[c]	0,56±0,1 5[b]	0,54±0,1 2[b]
Sudanese		T	58,1±11, 49[a]	3,35± 0,23[a]	0,95±0,0 1[a]	0,83±0,0 5[a]
	Glazing	A	31,9± 7,34[b]	2,62 ±0,19[b]	0,90±0,0 2[a]	0,76±0,0 5[a]
		B	25,0 ±4,71[b]	2,32 ±0,19[b]	0,85±0,0 4[a]	0,73±0,0 6[a]
		C	10,9 ±5 ,53[c]	1,27± 0,44[c]	0,57±0,1 9[b]	0,55±0,1 5[b]
		Pval ue				

Comparisons are made by column (diversity indices for each topographical unit are compared according to invasion level). Superscript letters indicate significant differences. Values with the same letters indicate no significant difference. Topo: topography

S: species richness; H: Shannon diversity index; D: Simpson diversity index; E: Piélou equitability; M±E: mean plus or minus standard deviation

3.8.3. Floristic similarity between invasion levels according to land use types

In the Sahelian zone, the herbaceous flora of pastoral zones and fallows varies significantly according to the level of invasion of *Senna obtusifolia*. Sørensen's similarity index shows average similarity between invasion levels B in fallow and pastoral areas, and high similarity between levels C in pastoral and fallow areas. However, low similarity was observed between the A and B levels of fallow land (Table 17).

In the Sudanian zone, Sørensen's similarity index shows similarity between invasion levels A and C in pastoral and fallow areas. There is little similarity between level A in the pastoral zone and B in the fallow, and between A and B in the fallow (Table 18).

Table 17 Sørensen's similarity index of the herbaceous stratum between *Senna obtusifolia* invasion levels in fallow and pastoral areas in the Sahelian zone

		Tp	Ap	Bp	Cp	Tj	Aj	Bj	Cj
Pastora l zone	Tp	1							
	Ap	0,422 8	1						

		Tp	Ap	Bp	Cp	Tj	Aj	Bj	Cj
	Bp	0,2014	0,4544	1					
	Cp	0,0468	0,2069	0,4506	1				
Fallow land	Tj	0,3281	0,3282	0,213	0,0468	1			
	Aj	0,1795	0,3601	0,3727	0,2094	0,3723	1		
	Bj	0,1174	0,3108	**0,5346**	0,4053	0,2373	**0,5726**	1	
	Cj	0,0248	0,1634	0,3621	**0,7463**	0,0437	0,2122	0,4306	1

Bold: similarity ≥ 5%, T: not invaded; A: lightly invaded, B: moderately invaded and C: heavily invaded. Zp: pastoral zone

Table 18 Sørensen's similarity index of the herbaceous stratum between *Senna obtusifolia* invasion levels in fallow and pastoral areas in the Sudanian zone

		Tp	Ap	Bp	Cp	Tj	Aj	Bj	Cj
Pastoral zone	Tp	1							
	Ap	0,398	1						
	Bp	0,2136	0,4871	1					
	Cp	0,1075	0,3173	0,4917	1				
Fallow land	Tj	0,4609	0,3453	0,2074	0,1129	1			
	Aj	0,3381	**0,5**	0,3976	0,2751	0,4072	1		
	Bj	0,3009	**0,5066**	0,4904	0,4022	0,325	**0,578**	1	
	Cj	0,1121	0,2948	0,4381	**0,5617**	0,1152	0,26	0,3735	1

Bold: similarity ≥ 5%, T: not invaded; A: lightly invaded, B: moderately invaded and C: heavily invaded. Zp: *pastoral* zone

3.8.4. Floristic similarity between invasion levels according to topography

In the Sahelian zone, the herbaceous stratum of lowlands and glacis varies significantly according to the level of invasion by *Senna obtusifolia*. Average similarity was observed between invasion levels A, B and C in lowlands and glacis. However, no similarity was observed between topographic level controls (Table 19). In the Sudanian zone, Sørensen's similarity index shows similarity between invasion levels T; A; B and C in lowlands and glacis (Table 20).

Table 19Sørensen's similarity index of the herbaceous stratum between invasion levels of *Senna obtusifolia* in lowlands and glacis in the Sahelian zone.

		Tb	Ab	Bb	Cb	Tg	Ag	Bg	Cg
Basfond	Tb	1							
	Ab	0,4535	1						
	Bb	0,3932	0,3982	1					
	Cb	0,4076	0,3497	0,4128	1				
Glazing	Tg	**0,5226**	0,4165	0,3307	0,2712	1			
	Ag	0,3221	**0,5736**	0,3945	0,3821	0,4634	1		
	Bg	0,1832	0,4831	**0,5103**	0,4532	0,4056	0,4764	1	
	Cg	0,1298	0,1935	0,4741	**0,5210**	0,1767	0,33	0,4132	1

Bold: similarity ≥ 5%, T: not invaded; A: slightly invaded, B: moderately invaded and C: heavily invaded

Table 20Sørensen's similarity index of the herbaceous stratum between invasion levels of *Senna obtusifolia* in lowlands and glacis in the Sudanian zone.

		Tb	Ab	Bb	Cb	Tg	Ag	Bg	Cg
Basfond	Tb	**1**							
	Ab	0,4654	1						
	Bb	0,3098	0,4523	1					
	Cb	0,2987	0,3276	0,4245	1				
Glazing	Tg	**0,5412**	0,3165	0,3934	0,2491	1			
	Ag	0,4076	0,4832	0,4298	0,3032	0,4172	1		
	Bg	0,2394	0,4205	**0,5402**	0,3521	0,3165	0,3754	1	
	Cg	0,1556	0,2110	0,4387	**0,5223**	0,1198	0,1843	0,0,3278	**1**

Bold: similarity ≥ 5%, T: not invaded; A: slightly invaded, B: moderately invaded and C: heavily invaded.

3.9. Dominant herbaceous species by climatic zone and level of invasion

A total of 37 herbaceous species are dominant according to their relative importance value ≥ 5. In the Sahelian zone, the most dominant are those with the highest relative importance value in invasion level C, namely *Senna obtusifolia (IR= 98.01); Eragrostris tremula* (IR=13.23*); Panicum laetum* (IR=11.58); *Digitaria horizontalis* (IR=8.06); *Dactyloctenium aegyptium* (IR=6.28); *Chrysanthellum americanum* (IR=5.17); *Spermaceae stachydea* (5.17) and *Desmodium hirtum* with an IR of 5.07 (Table 21). These species are of low pastoral value. In the Sudanian zone, *Senna obtusifolia* (IR=9.47), *Hyptis suaveolens* (IR=7.58) and *Sida rhombifolia* (IR=5.21*)* are the most dominant herbaceous species and are either of low or no pastoral value.

Table 21 Dominant herbaceous species (relative importance ≥ 5) climatic zones according to *Senna obtusifolia* invasion levels

Species	VP	Sahel				Sudanese			
		T	A	B	C	T	A	B	C
Achyranthes aspera	PLEASE	2,18	1,43	2,59	2,60	2,01	2,77	3,77	3,32
Alysicarpus ovalifolius	BVP	2,89	4,83	4,15	0,95	—	—	—	—
Aristida adscensionis	BVP	2,02	2,81	0,71	—	—	—	—	—
Brachiaria lata	BVP	2,42	2,85	1,62	2,73	—	—	—	—
Chamaecrista mimosoides	MVP	3,09	4,03	4,21	0,83	—	—	—	—
Chloris gayana	BVP	3,02	3,75	2,77	1,63	—	—	—	—
Chloris pilosa	BVP	5,07	6,30	7,00	4,25	—	—	—	—
Chrysanthellum americanum	PLEASE	2,17	2,85	3,43	**5,17**	—	—	—	—

72

Species	VP	Sahel				Sudanese			
		T	A	B	C	T	A	B	C
Corchorus olitorius	MVP	2,20	1,50	1,17	0,83	1,48	2,44	2,22	3,32
Corchorus tridens	MVP	2,20	3,37	3,95	4,38				
Cyperus difformis	FVP	—	—	—	—	1,59	1,30	1,77	2,00
Cyperus iria	FVP	3,34	3,62	2,07	0,95				
Dactyloctenium aegyptium	MVP	3,60	7,86	6,87	6,28	1,27	1,47	1,99	3,32
Desmodium hirtum	MVP	2,34	5,32	4,85	5,07				
Digitaria horizontalis	MVP	3,94	6,07	5,86	8,06	1,27	2,28	2,44	2,84
Echinochloa colona	MVP	4,24	2,78	2,99	3,68	—	—	—	—
Eragrostis tremula	FVP	6,19	15,26	17,37	**13,23**	—	—	—	—
Gomphrena celosioides	FVP	—	—	—	—	1,38	1,95	1,99	2,00
Hyptis suaveolens	PLEASE	—	—	—	—	1,91	3,09	3,77	**7,58**
Indigofera bracteolata	MVP	—	—	—	—	0,53	0,98	0,89	2,37
Kyllinga pumila	PLEASE	—	—	—	—	1,48	2,12	2,44	2,37
Mitracarpus scaber	FVP	—	—	—	—	2,01	3,09	3,10	3,32
Ocimum canum	PLEASE	—	—	—	—	1,17	1,63	1,99	2,37

Species	VP	Sahel				Sudanese			
		T	A	B	C	T	A	B	C
Panicum laetum	FVP	7,55	17,82	15,58	**11,58**	—	—	—	—
Pennisetum pedicellatum	BVP	—	—	—	—	2,00	2,44	2,66	1,89
Schoenefeldia gracilis	MVP	3,14	2,71	3,57	0,83	—	—	—	—
Senna obtusifolia	FVP	—	**22,70**	**48,32**	98,01	—	**3,26**	**4,43**	**9,47**
Setaria pumila	BVP	4,01	5,56	3,79	1,65				
AIDS acuta	PLEASE	—	—	—	—	1,91	2,61	2,88	3,79
Sida rhombifolia	PLEASE	—	—	—	—	1,59	2,28	2,44	**5,21**
AIDS urens	BVP	—	—	—	—	1,70	2,28	2,66	3,32
Spermacoce radiata	PLEASE	2,93	**8,38**	**6,06**	1,65	—	—	—	—
Spermacoce stachydea	PLEASE	4,15	11,55	11,77	5,17	—	—	—	—
Spermacoce verticillata	PLEASE	—	—	—	—	1,59	1,47	1,77	2,37
Tephrosia pedicellata	PLEASE	—	—	—	—	1,27	1,95	1,99	2,00
Triumfetta rhomboidea	PLEASE	—	—	—	—	1,70	2,44	2,88	3,32

		Sahel				Sudanese			
Species	VP	T	A	B	C	T	A	B	C
Zornia glochidiata	BVP	4,41	**8,42**	**6,72**	2,73	—	—	—	—

Legend: BVP: Good Pastoral Value; MVP: Average Pastoral Value; FVP: Low Pastoral Value; SVP: No Pastoral Value

Dashes (-) signify a relative importance value of < 2 in all invasion levels.

3.10. Herbaceous biomass by level of invasion in climatic zones

Analyses reveal that in the Sahelian climatic zone, the level of invasion of *S. obtusifolia* has a highly significant effect (ddl. =3; F=21.17; P < 0.0001) on the biomass of other species in the herbaceous stratum. Indeed, the biomass of other species is highest (2.34 ± 1.09 t/ha) when the invasion rate is between 5% and 35%, and very low (0.82 ± 0.52 t/ha), for an invasion rate greater than 65%. In contrast, the above-ground biomass of *S. obtusifolia* increases with the rate of invasion, ranging from 0.64 ± 0.54 t/ha at the lowest level of invasion to 4.47 ± 1.99 t/ha at the highest (Table 22).

Also in the Sudanian climatic zone, the level of invasion by *S. obtusifolia* had a highly significant effect (ddl. =3; F=57.17; P < 0.0001) on the biomass of other herbaceous species. This biomass decreases in invaded environments compared to the control. It is highest with a mean value of 3.17 ± 1.34 t/ha for invasion rates between 5% and 35%, and lowest when the invasion rate exceeds 65%, with a mean value of 1.06 ± 0.51 t/ha (Table 23). The above-ground biomass of *S. obtusifolia* increases progressively with the level of invasion. It is 1.31 ± 1.46 t/ha for the lowest level and 5.83 ± 3.21 t/ha for the highest (Table 23).

Table 22 Variation in above-ground biomass *of Senna obtusifolia* and other species by level of invasion in the Sahelian zone

	Invasion levels of *S.* obtusifolia					
	T	A	B	C	F	P
Above-ground biomass of *S. obtusifolia*	-	0,64±0,53 [a]	1,31±0,72 [b]	4,47 ± 1,9[c]	156,8	<0,001
Above-ground biomass of other	2,07±1,18 [a]	2,33±1,09 [b]	1,87±1,44 [c]	0,83± 0,51[d]	21,17	<0,001

herbaceous plants		

Levels not connected by the same letter are significantly different at the 5% level.

Legend: T: non-invaded level; A: slightly invaded level; B: moderately invaded level; C: heavily invaded level;

Table 23 Variation in above-ground biomass *of Senna obtusifolia* and other species in the Sudanian zone

	\multicolumn{6}{c}{Invasion levels of *S.* obtusifolia}					
	T	A	B	C	F	P
Above-ground biomass of *S. obtusifolia*	-	1,3 ±1,46[a]	1,97±1,14[b]	5,83±3,21[c]	78,3 6	<0,00 1
Above-ground biomass of other herbaceous plants	4,13±2,08[a]	3,17±1,3 4[b]	2,26± 0,89[c] .	1,06±0,5 1[d]	57,1 2	<0,00 1

Levels not connected by the same letter are significantly different at the 5% threshold.

Legend: T: non-invaded level; A: slightly invaded level; B: moderately invaded level; C: heavily invaded level;

3.11. Pastoral value as a function of *Senna obtusifolia* invasion levels

In the Sahel, the graphical representation of pastoral value shows that the rate of recovery of species of Low Pastoral Value (LPV) increases with the level of invasion by *S. obtusifolia* (Figure 14). At high levels of invasion (over 65%), recovery is maximal, reaching 100%. Among these Low Pastoral Value (LPV) species, *S. obtusifolia* has a cover rate of 99%, compared with 1% for the other species in level C. The coverage of species with Medium Pastoral Value (MVP) and Good Pastoral Value (BVP) is higher in the control or lightly invaded levels (invasion level A), and virtually nil in the heavily invaded levels (invasion levels B and C). We also record a maximum cover rate (over 10%) of species of no forage value (SVP) in the control levels. This rate decreases as the level of invasion increases.

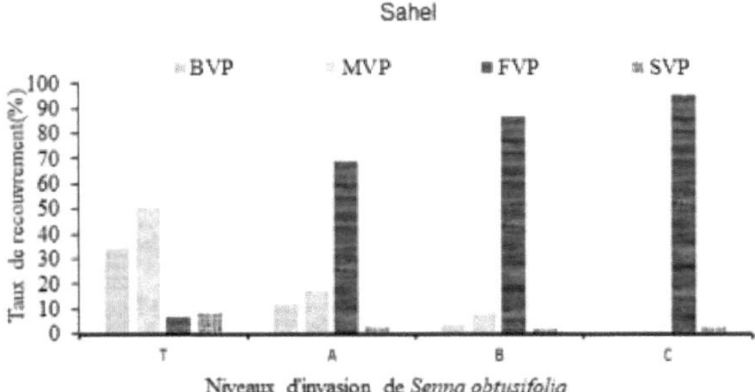

Figure 14Graphical representation of pastoral value as a function of *S. obtusifolia* invasion level in the Sahelian zone.

SVP: species with No Pastoral Value; FVP: species with Low Pastoral Value; MVP: species with Medium Pastoral Value; BVP: species with Good Pastoral Value.

In the Sudanian zone, the results observed are almost similar to those in the Sahelian zone, where the cover rate of low forage value species increases with the invasion of *S. obtusifolia* (Figure 15). Invasion level C (level above 65%) comprises 90% of Low Value species versus 10% of species with no forage value. Among the species with Low Pastoral Value, *S. obtusifolia* has a cover rate of 80%, compared with 20% for the other species.

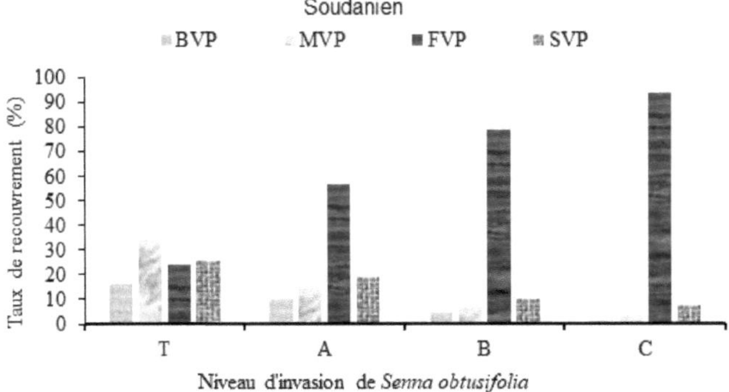

Figure 15Graphical representation of pastoral value as a function of *Senna obtusifolia* invasion level in the Sudanian zone.

SVP: species with No Pastoral Value; FVP: species with Low Pastoral Value; MVP: species with Medium Pastoral Value; BVP: species with Good Pastoral Value.

4. Discussion

4.1 Influence of environmental factors on the invasiveness of *Senna obtusifolia*

Climatic zone, type of land use, topography and their interaction had a significant effect on the invasion of *S. obtusifolia*, expressed here in terms of growth (height), recruitment (density) and development (biomass). Indeed, in the Sahelian zone, the density of the species is higher than in the Sudanian zone. The Sahel being a livestock zone (Ayatunde *et al.*, 2020) with less rainfall (from 400mm to 600mm/year) is prone to pockets of prolonged drought, leading to a reduction in forage resources. This scarcity of fodder forces livestock to consume the dry pods of *S. obtusifolia* during this dry period, and to disseminate the seeds during their grazing. Our results corroborate those of Thiombiano *et al.* (2012) and Ouédraogo *et al.* (2021), who found that cattle promote the spread of *S. obtusifolia*. The density of *S. obtusifolia* is higher in pastoral areas than in fallow land. This high density of *S. obtusifolia* is thought to be due to the high frequency of cattle in these areas. This enables them to consume the dry pods of *S. obtusifolia* and spread the seeds, especially in the dry season. Kiéma *et al* (2012) found that animals consume the dry pods of *S. obtusifolia* and spread the seeds contained in their droppings during grazing.

Topography has a significant effect on the proliferation of *S. obtusifolia*. The species is more abundant with large size and high biomass in lowlands than in glacis. This may be partly explained by the fact that run-off water carries *S. obtusifolia* seeds into the shallows. Grubben and Denton (2004) and Kadéba *et al.* (2015) found that *S. obtusifolia* is very abundant in low-lying areas along rivers. On the other hand, moisture in the shallows would have favored the large size as well as the high biomass of individuals of the species compared to the glacis where the size and biomass of *S. obtusifolia* are reduced.

Individuals selected for height and yield parameter measurements showed greater height, high number of pods per plant, higher average pod weight, number of seeds per pod and seed weight in the Sudanian zone than in the Sahelian zone. This could be explained by the fact that increased rainfall allows greater availability of nutrients and water (Santiago *et al.*, 2004), which is favorable to plant development. A strong positive correlation was reported between average pod weight and the weight of seeds contained in the pod. This suggests that the seeds occupy a large portion of the pod.

4.2 Impact of the *Senna obtusifolia* invasion on phytodiversity

The heavy invasion of the study sites by *S. obtusifolia* has led to the regression of several species, and thus to a gradual reduction in biodiversity. It is in this sense that Muller (2000) asserts that invasive species can have a negative but very subtle impact on biodiversity or reflect ecosystem dysfunction. As a reminder, the species inventoried are largely dominated by therophytes. This dominance of annual species reflects the level of disturbance of the habitats colonized by *S. obtusifolia*. This implies that these therophytes have established themselves after a series of disturbances that have freed up space for their benefit (Goudard, 2007). Generally speaking, the climate-influenced invasion of *S. obtusifolia* has affected the number and distribution of plant species in both climatic zones. In the Sahelian zone, climatic conditions appear to be more favorable to the establishment of *S. obtusifolia*, to the detriment of other herbaceous species. On the other hand, in the Sudanian climatic zone, climatic conditions are more favorable to the establishment of various plant species. These results are in agreement with those of Zerbo (2018), who found that species richness increases with increasing rainfall and rainy season duration. So in the absence of anthropogenic pressure and in habitats not yet dominated by the invasive species, floristic diversity is high. In terms of land use, pastures invaded by *S. obtusifolia* are home to fewer species than less invaded pastures such as fallow land. Pastures are permanently frequented by livestock, which more often than not causes overgrazing, unlike fallow land, which is only occasionally frequented by livestock. Similar results were reported by Idrissou *et al* (2020), who assert that taxonomic richness (number of species, families and genera) decreases as grazing intensity increases. Akpo and Grouzis (2000) also reported that high grazing intensity leads to a decrease in the floristic richness of pastures. Diversity decreases as the level of invasion increases. In fact, overgrazing has led to a frequent selection of palatable species in favor of less palatable or unpalatable species, which have gained more space. Mechanisms such as herbivory have been identified as factors that can predominantly explain susceptibility to invasion (Levine *et al.*, 2004). Invasion levels (C) are similar in fallow and pastoral areas. These levels are dominated by a high density of *Senna obtusifolia* and the virtual disappearance of other species. It has been reported that when the environment is favorable to its establishment, *Senna obtusifolia* is able to exclude all native species and form mono-specific environments, causing ecological change (Mackey *et al.*,

1997; Tye *et al.*, 2003). Invasive species can eliminate less competitive, but often highly useful, rare or even endemic species (Muller, 2000). Some invasive species can even profoundly disrupt the natural habitat, making it unsuitable for local species (Lowe *et al.*, 2007). Gebrekiros and Tessema (2018) revealed that floristic diversity is high in sites not or slightly invaded *by S. obtusifolia*. On the other hand, the strong interspecific competition established by the invasive species in heavily invaded habitats is not conducive to the establishment of a high diversity of plant species.

The invasion of *S. obtusifolia* reduces the productivity of other species in the herbaceous stratum. Even in lightly invaded habitats, the biomass produced by *S. obtusifolia* exceeds a quarter of the total biomass of the herbaceous stratum. With an invasion rate of over 65%, the biomass of *S. obtusifolia* exceeds 70% of the total biomass of the herbaceous stratum. In this case, the biomass of the other species in the herbaceous stratum is reduced by more than half compared with the potential biomass. These results are in line with those of Gebrekiros and Tessema (2018), who reveal low production of other herbaceous species in environments invaded by *S. obtusifolia*. As this species is not very palatable when fresh (Kiema *et al.,* 2012), grazing pressure will be higher in control and lightly invaded levels. Biomass production in natural savannah varies greatly according to grazing pressure (Botoni *et al.,* 2006). In a grazed area, phytomass can represent 50 to 90% of its potential value (Fournier, 1994). The invasion of *S. obtusifolia* is reflected in its proliferation and increased growth. The species monopolizes the nutritive resources of the environment at the expense of other species, which, under the influence of this interspecific competition, see their development reduced. According to Gebrekiros and Tessema (2018), *S. obtusifolia* monopolizes light, moisture, nutrients and space to the detriment of other species. Indeed, during data collection, observations revealed that in invasion levels C (invasion $\geq$ 65%), the other herbaceous species present a habit, rather elongated, less spreading and very often of frail appearance.

The cover rate of species with low pastoral value increases with the invasion of *S. obtusifolia*. Invasion translates into a proliferation of individuals of the species, and therefore an increase in their rate of cover. Indeed, *S. obtusifolia* itself is a species of low pastoral value (Kiéma, 2002; 2007; 2008; Sawadogo, 2011), with a strong presence in invaded environments. As a result, the evolution of its cover rate is indicative of the cover rate of species of low pastoral value in an environment. From an invasion rate of 65%, species of low pastoral value occupy virtually the entire surface of the environment, with *S. obtusifolia* alone occupying 90% of this surface. In both climatic zones, *S. obtusifolia* is the species with the highest relative importance value. These values revealed a clear dominance of species with little or no pastoral value. After *S. obtusifolia* comes *Eragrostris tremula* in the Sahelian zone and *Hyptis suaveolens* in the Sudanian zone. The coverage rate of species with Medium and Good Pastoral Values is higher in control levels and in lightly invaded environments than in heavily invaded ones. The invasion of *S. obtusifolia* therefore contributes to reducing the availability of moderately and highly palatable species. These results are in line with those of Gebrekiros and Tessema (2018), who found that the rate of rejects or unpalatable species is high in sites invaded by *S. obtusifolia*. Floristic composition is the essential determinant of pasture quality (Kiéma, 2002). As a result, pasture quality, and hence extensive livestock farming, is directly threatened by the invasion of *S. obtusifolia*. However, species with no pastoral value dominate the invaded levels (over 55% cover). This could be explained by the high selectivity of the species eaten during grazing. The result is an increase in the refusal rate. Several authors (Guinko, 1984; Botoni *et al.*, 2006; Alhassane *et al.*, 2018) have found that high pastoral pressure and high selectivity of palatable species can lead to the degradation of pastoral flora by paving the way for the establishment of species with no pastoral value, indicative of rangeland depletion.

5. Partial conclusion

This activity shows that climate, land use, topography and their interaction influence the proliferation of *S. obtusifolia*. The first hypothesis, that climate is the most decisive factor in the invasion of *S. obtusifolia*, is invalidated. In the Sahelian zone, the density of *S. obtusifolia* is higher than in the Sudanian zone, where a low density of the species is observed. However, the biomass of *S. obtusifolia* is lower in the Sahelian zone than in the Sudanian zone. Pastures are habitats characterized by a greater proliferation of *S. obtusifolia* than fallow land. This proliferation of *S. obtusifolia* is subject to its dissemination, but its growth is better in fallows, which are reconstituting soils. Topography has a significant effect on *S. obtusifolia* proliferation. The species is more abundant with a large size and high biomass in lowlands than in glacis. The second hypothesis, that topography influences the yield parameters of *S. obtusifolia*, is confirmed. This activity reveals that the invasion of *S. obtusifolia* modifies the species richness of herbaceous and juvenile woody plants in both climatic zones, thus contributing to a reduction in floristic diversity. The third hypothesis, which states that taxonomic richness decreases with increasing levels of *S. obtusifolia* invasion, is confirmed. The forage quality of invaded pastoral zones decreases with the increase in the cover rate of species with Low Pastoral Value and the regression of that of species with Medium and Good Pastoral Value. The fourth hypothesis, which states that pastoral value decreases with increasing levels of *S. obtusifolia* invasion, is confirmed. A high level of *S. obtusifolia* invasion leads to a drastic drop in the above-ground biomass produced by the rest of the herbaceous stratum. However, pasture degradation cannot be attributed solely to invasion by *S. obtusifolia*. Changes in pasture flora and vegetation at any given time are the result of the combined action of several anthropic and ecological factors (animals, climate). However, it is important to ask whether the rapid invasion of *S. obtusifolia* is due in part to its ability to dominate other herbaceous plants?

CHAPTER IV: COMPETITIVENESS OF *SENNA OBTUSIFOLIA* (L.) H. S. IRWIN & BARNEBY AGAINST THREE HERBACEOUS SPECIES UNDER DIFFERENT WATER REGIMES

From this chapter, an article was drawn and published under the following reference:

Alhassane Zaré, Pawend-taore Christian Bougma, Ouedraogo Karim, Oumarou Ouédraogo (2022): Assessing the competitive ability of the invader *Senna obtusifolia* with coexisting native species under different water stress regimes. *Journal of Experimental Biology and Agricultural Sciences*. DOI: http://dx.doi.org/10.18006/2022.10(5).1149.1167

1. Introduction

The success of invasive plants depends on several abiotic and biotic factors (Martin & Coetzee, 2014; Leal *et al.*, 2022) as well as biological traits linked to their colonization capacity (Wickert *et al.*, 2017; Yu *et al.*, 2018; El-Barougy *et al.*, 2020). In general, abiotic factors such as water, light and nutrients for which plants compete in the growth and development process (Wang *et al.*, 2012) can influence the intensity of competition (Yu *et al.*, 2018) when these resources are limited between close individuals with a common need (Bastiani *et al.*, 2016; Zhou *et al.*, 2018). Current climatic conditions could be one of the causes of this invasion. Climate forecasts show a downward trend in precipitation, leading to more intense droughts. This could also facilitate the proliferation of invasive plants (Diez *et al.*, 2012; Paudel *et al.*, 2018). For example, in the Sahelian pastures of West Africa, the arid climate characterized by severe drought (Issaharou-Matchi *et al.*, 2016) has favored the proliferation of invasive plants due to their high physiological performance (Oliveira *et al.*, 2014). In this situation, the species with a high capacity in resource acquisition can overcome its congeners and become invasive (Funk *et al.*, 2016; Broadbent *et al.*, 2018). *Senna obtusifolia* has been identified as a herbaceous species that has invaded Sahelian pastures. Its invasion therefore represents a major challenge for livestock farming, heightening the problem of the availability of quality fodder for livestock.

The objectives of this study are to :

- to determine the colonization capacity of *S. obtusifolia* in a controlled environment;
- test the sensitivity of *S. obtusifolia* to water deficit in a controlled environment.

The assumptions underlying this activity stipulate that:

- the competitive ability of *S. obtusifolia* is due to its biological characteristics;
- water deficit is a factor that improves the performance of *S. obtusifolia to the* detriment of its congeners.

2. Presentation of the study site

The experiment was carried out at the Centre National des Semences Forestières (CNSF) in Ouagadougou, Burkina Faso. It is located north of the city of Ouagadougou in Burkina Faso, on the Ouagadougou-Kaya axis between 12°24'31.78"N and 01°28'59.33" W (Figure 16). The central region is located in the Sudano-Sahelian climatic zone (Guinko, 1984). The climate is characterized by a relatively long dry season from October to May and a rainy season from May to mid-October, with very intense rains in July and August. The average annual rainfall in the Sudano-Sahelian zone is between 600 and 900 mm. Climatic data for the study area (Ouagadougou) obtained from the Agence Nationale de la Météorologie du Burkina show an average annual rainfall of around 860 mm, with an average annual temperature of 28.86°C in 2018. The CNSF's average temperature and humidity were 30.2 to 32.54°C and 62.46 to 74.47% respectively from July to October 2019.

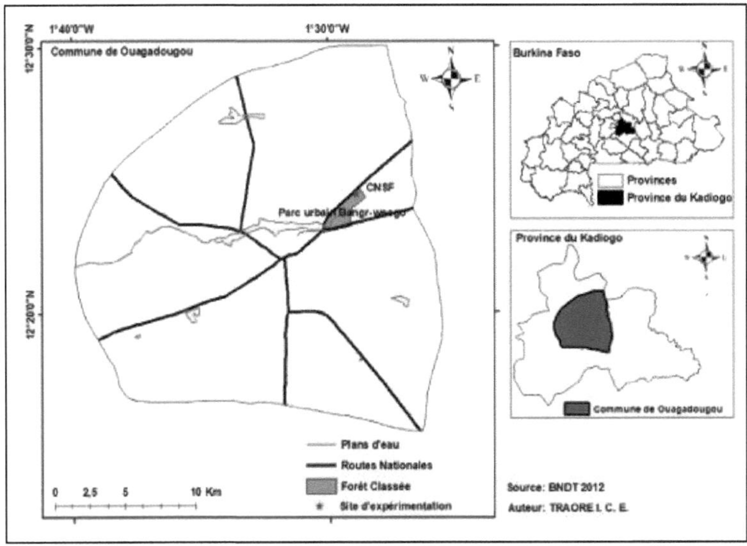

Figure 16 Study area overview

3. Description and rationale for choice of species in competition with *Senna obtusifolia*

Three species with a high forage value and belonging to different functional groups were selected for experimentation. They are : *Andropogon gayanus* (perennial grass), *Pennisetum pedicellatum* (annual grass) and *Chamaecrista mimosoides* (annual legume). These three species also have a high capacity for drought resistance (Adams & Setterfield, 2013; Issaharou-Matchi *et al.*, 2016) and each was combined with *S. obtusifolia* in greenhouse pots under water deficit. Indeed, the competitive capacity between legumes and grasses for resources has already been tested by several authors (DeBoer *et al.*, 2020; Grüner *et al.*, 2020; Khatiwada *et al.*, 2020).

3.1. Biology and ecology of s species in competition with *Senna obtusifolia*

3.1.1. *Chamaecrista mimosoides* (L.)

Generally an annual, it can take on a perennial shrub form in humid regions (Berhaut, 1967). *C. mimosoides* belongs to the Fabaceae family. An upright plant, 30 to 70 cm high, up to 120 cm, weakly branched with an elliptical profile, *C. mimosoides* has alternate, compound, paripinnate leaves. The flowers are yellow, isolated to fasciculated by 2 to 4 and inserted clearly above the petiolate axil, on the stem (Photo 5). The fruits **of** this plant are linear, flattened, very slightly arched and pubescent pods. Each pod averages 5 cm long and 0.5 cm wide. The pod contains 12 to 24 seeds arranged in a single series. The valves roll up after dehiscence. Seeds are small and flat, rectangular in shape. They are 3.5 mm long, 2 mm wide and 0.6 mm thick. They have a smooth, light-beige seed coat. It reproduces only by seed. The development cycle is relatively long. Germination takes place in the wet season after heavy rains. Flowering doesn't occur until the cool season (August) and fruiting lasts until the end of the dry season (Figuiere *et al.,* 1998). This leguminous plant also has a low forage value. It can produce nodules that fix soil nitrogen, especially under extreme conditions (high humidity or drought). This species of the *Chamaecrista* genus contains toxic substances that can cause muscular and neurological damage when consumed by livestock in large doses (Berhaut, 1967).

Photo 4*Chamaecrista mimosoides*, author: ZARE Alhassane CNSF 09/2019

Chamaecrista mimosoides (L.) is widespread throughout tropical Africa and Asia. It is also found in Australia. It thrives on sandy, well-drained soils, where it forms scattered populations that do not present a high potential for invasion. While not a highly troublesome weed, *C. mimosoides* is not classified as an invasive species (CBD, 2014), but its status differs from country to country. In Niger, it is potentially invasive according to CNEDD (2010).

3.1.2. *Andropogon gayanus* Kunth.

Andropogon gayanus Kunth. is a perennial grass belonging to the Poaceae family. It is a large, rough, upright perennial plant, growing from 1 to 3 m high, and can even exceed 4 m. The stem has internodes and is entirely covered by leaf sheaths during growth. The stem has internodes and is entirely covered by leaf sheaths during the growth phase. At this stage, the plant forms numerous thalli. At bolting, each of these tillers can develop a long, soft stem bearing more than a dozen leaves and ending in a false panicle inflorescence around 40 to 60 cm long. With a short rhizome of inter-nodes and intra-vaginal branching, it forms a clump around 1 m in diameter thanks to abundant tillering (Photo 6) according to Traoré (1996). Unlike annuals, which survive the dry season as seeds, perennial grasses survive the dry season as vegetative plants. The presence of residual vegetative parts and a radical apparatus enables them to regrow rapidly when the first rains come. The complete seasonal development cycle lasts an average of 6 to 7 months, depending on ecotype and geographical location. It lasts 4 months in the Sahelian zone and reaches 9 months in the Guinean zone (Traoré, 1996).

Photo 5*Andropogon gayanus* clumps; author
ZARE Alhassane Sidéradougou 10/2018

Andropogon gayanus Kunth. is native to African savannas, where it is particularly prolific in fallow land (Fournier, 1994), and is widespread in most African tropical savannas between isohyets of 400 and 1500 mm (Traoré, 1996). Where soil and topographical conditions are favorable, the species can thrive in rainfall of less than 400 min. This species can be found in a wide variety of environments, including fields, fallow land and the edges of embankments, tracks and roads. In Burkina Faso, the species is found in savannahs, with a more pronounced distribution in the Sudano-Sahelian zone (Thiombiano *et al.,* 2010). It is used as a forage species to restore degraded pastures and appears to be a grass that competes effectively with invasive species (Mitja *et al.,* 2004).

3.1.3. *Pennisetum pedicellatum* Trin.

Pennisetum pedicellatum belongs to the Poaceae family. It *is* a bushy annual grass with erect or spreading stems reaching 40 to 100 cm or more. The plant is generally very tall. The stem is cylindrical, branched and sometimes bears stilt roots. Leaves are linear to broadly linear, around 30 cm long and 1 cm wide; they are very broad in the middle and narrow at the tips, becoming pointed at the top and narrowing at the base.

The inflorescence is a highly contracted cylindrical panicle, looking like a terminal spike, 5 to 15 cm long and 1 to 2 cm wide (Photo 7), according to Poilecot (1999).

Photo 6Clumps of *Pennisetum pedicellatum,* author ZARE Alhassane DORI 9/2019

Pennisetum pedicellatum is found throughout the African tropical savannah. In the Sudanian zone, it is often associated with *Andropogon gayanus, Loudetia togoensis and Digitaria gayana* (Poilecot, 1999). The species grows well on deep and moderately deep soils. The development cycle lengthens with rainfall. The shorter the rainy season, the longer the cycles. Thus, cycles of 102 to 112 days are found at isohyets 600 mm, 120 days at 900 mm and 130 days at isohyets 1250 mm (Traoré 1996). For propagation, the species can be successfully sown naturally, in rows or broadcast. Under certain conditions, cuttings are also possible, especially when the soil is wet (Penning *et al.,* 1982). It is a species of good fodder value used to restore degraded pastures.

4. Methodology

4.1. Plant material

The plant material used to test the competitiveness of *S. obtusifolia* consisted of seeds (Photo 8) from two legumes (*S. obtusifolia* and *C. mimosoides*) and two grasses (*A. gayanus* and *P. pedicellatum*). The seeds of these species come from pastures in the Sahelian zone of Burkina Faso. This zone is considered Burkina Faso's leading livestock region (Ouedraogo and Ayantunde, 2019), where *S. obtusifolia* is one of the most widespread species (Zerbo *et al.,* 2016) and may be more tolerant of water deficit.

Indeed, the climate is Sahelian, with annual rainfall below 600 mm. The dry pods containing the seeds were harvested at maturity in mid-November directly from the plants to avoid infestation on contact with the soil, and placed in envelopes. Seeds harvested in November were stored in the laboratory at room temperature (25-30°C) and sown in June.

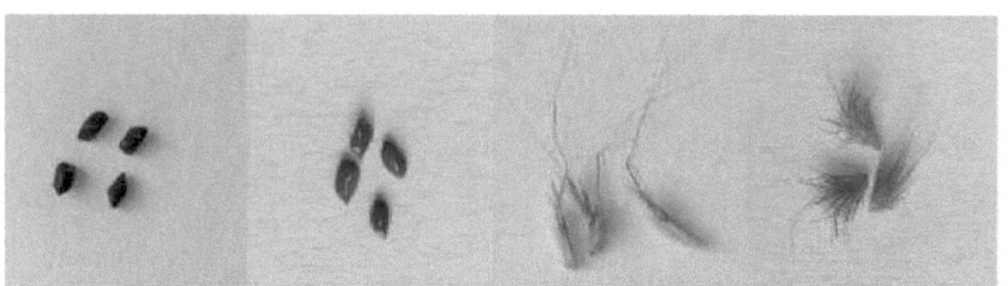

S. obtusifolia C. mimosoides A. gayanus P. pedicellatum

Photo 7Seeds of the species studied

4.2. Laboratory germination test

The laboratory germination test was carried out to determine the best germination rates. To do this, the water content of the seeds was assessed before they were germinated using a device.

4.2.1. Evaluation of seed water content

This test verifies seed quality before germination (Appendix 6). The loss of water, depending on the species, means that the seeds are better preserved. This could guarantee a higher germination capacity. By definition, according to ISTA (2009), the water content of a sample is its weight loss when dried according to these rules. Moisture content is expressed as a percentage of the initial sample weight.

For each species, four replicates of five seeds were used to assess moisture content. After weighing the empty cups and then the cups containing the seeds, the seeds were dried for 17 hours in an oven in accordance with ISTA (2009) regulations. After drying, the cups (the container containing the seeds) were cooled for 45 minutes in a desiccator before being weighed again. The water content was then determined according to the following formula:

$$TE = \frac{Pom - pim}{po\ m} \times 100$$

With : **P0m**= average fresh seed weight; **Pim**: average dry seed weight; **TE**: water content.

4.2.2. Experimental set-up in the laboratory

The laboratory germination test follows a two-block set-up, with one block containing seeds that have undergone pre-treatment (acid or water treatment) and the other containing control seeds, i.e. seeds that have not been pre-treated (Figure 17).

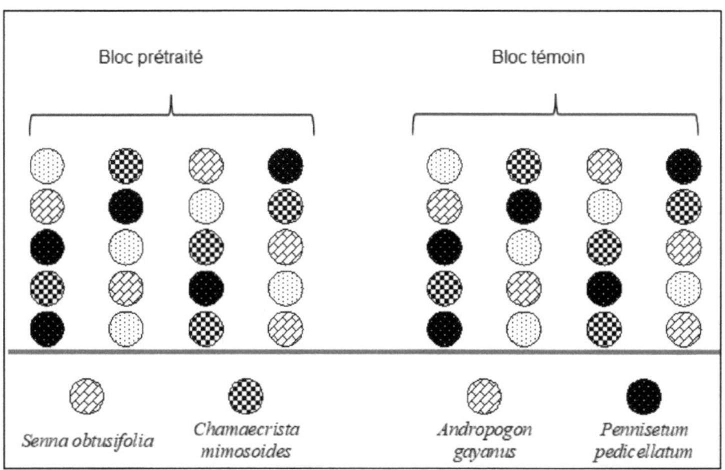

Figure 17Experimental set-up for germination test in the laboratory

4.2.3. Seed pre-treatment

These pre-treatments help to reduce germination time by rapidly breaking seed dormancy. This also helps to achieve the high germination rate required for the competition trial.

A first block of seeds was pre-treated with 96% concentrated sulfuric acid ($H_2 SO_4$) and water, and a second block with control seeds. Pre-treatment differed according to seed type. According to the CNSF forest seed catalog (2012-2015), pretreatment with 96% concentrated sulfuric acid ($H_2 SO_4$) is recommended for *S. obtusifolia* and *Chamaecrista mimosoides* seeds. For Andropogon *gayanus* and *Pennisetum pedicellatum* seeds, pre-treatment with water is suitable. Pre-treatment with concentrated sulfuric acid involved soaking *S. obtusifolia* and *C. mimosoides* seeds for 5 and 3 minutes respectively in a 96% solution of concentrated sulfuric acid ($H_2 SO_4$

). Water pre-treatment involves soaking in water for 24 hours. In each block, each batch of 25 seeds was repeated 5 times.

4.2.4. Germination

Untreated (control) and treated seeds of each species were germinated 24 hours after pre-treatment in Petri dishes with 5 replicates per treatment, at a rate of 25 seeds per dish at room temperature. The substrate used was sand sterilized at 150°C for one hour, and the seedlings were watered every 2 days using a watering can. The duration of the test was set at the germination period, which lasted from 8 to 16 days after sowing, during which virtually no further germination was observed.

4.3. Greenhouse trial

Seed germination activity in the laboratory was carried out to confirm seed quality through water content. Pre-treatments were also used to improve germination rates for each species, and to guide germination trials in the greenhouse.

4.3.1. Preparation of culture substrates

The substrate used consisted of a mixture of loamy soil, sand and manure at a ratio of 3:2:1, i.e. three (03) barrows of topsoil, two (02) barrows of sand and one (01) barrow of manure. This mixture was sterilized and distributed in plastic basins (top diameter: 50 cm, bottom diameter: 30 cm and depth: 25 cm) or pots, at a rate of 30 kg of substrate per basin.

4.3.2. Greenhouse experimental set-up

The experimental set-up used is a split plot consisting of four blocks and two treatments (water regime and species association). Sowing took place on June 22, 2019 in the various pots. In each pot, 25 seeds of each species were sown. Depending on the combinations, the number of seeds per pot was 25 seeds for the control pots, i.e. a single species, and 50 seeds per pot for the two-species combinations. Each block was watered with a quantity of water determined from field capacity. Each block contains 25 pots of monoculture and polyculture species (two-species combinations at different densities). Each pot was repeated five times to minimize experimental errors (Figure 18). Species were planted in association or not in each pot, including :

- A control is represented by a single species (*S. obtusifolia* or *C. mimosoides* or *A. gayanus* or *P. pedicellatum*).

- An association of two species: *S. obtusifolia* and another species (*S. obtusifolia* + *C. mimosoides*; *S. obtusifolia* + *A. gayanus*; *S. obtusifolia* + *P. pedicellatum*).

In this experiment, the proportions of each species in monoculture varied from pot to pot, ranging from 3; 6; 9 to 12 species on the one hand. On the other hand, *S. obtusifolia* was grown in association with each of these three species: *A. gayanus, C. mimosoides* and *P. pedicellatum*. For each association, the proportions are as follows: 12:0; 9:3; 6:6; 3:9 and 0:12, as shown in Figure 19. A total of 25 pots were considered, i.e. 16 pots in monoculture and 9 pots in mixed culture. Each pot was replicated five times, giving a total of 80 pots in monoculture and 45 pots in mixed culture, or 125 pots per block.

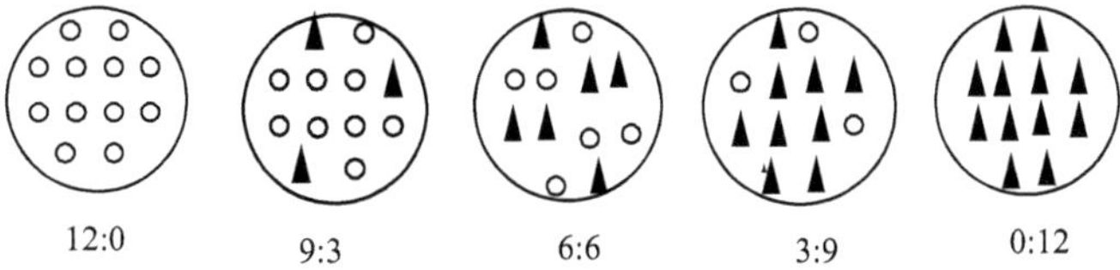

Figure 18Schematic diagram of the replacement series model (O circle represents *Senna obtusifolia* and ▲ triangle represents one of the other 3 species).

Caption:

Monoculture

S. obtusifolia (3 individuals) ; *S. obtusifolia* (6 individuals) ; *S. obtusifolia* (9 individuals) *S. obtusifolia* (12 individuals) ; *C. mimosoides* (3 individuals) ; *C. mimosoides* (6 individuals)

C. mimosoides (9 individuals) ; *C. mimosoides* (12 individuals) ; *A. gayanus* (3 individuals) ; *A. gayanus* (6 individuals) ; *A. gayanus* (9 individuals) *A. gayanus* (12 individuals)

P. pedicellatum (3 individuals); *P. pedicellatum* (6 individuals); *P. pedicellatum* (9 individuals)

P. pedicellatum (12 individuals)

Species association

Association of *S. obtusifolia* + *C. mimosoides* (3 :9) ; Association of *S. obtusifolia* + *C. mimosoides* (6 :6) ; Association of *S. obtusifolia* + *C. mimosoides* (9 :3) ; Association of *S. obtusifolia* + *A. gayanus* (3 :9) ; Association of *S. obtusifolia* + *A. gayanus* (6 :6) ; Association of *S. obtusifolia* + P. pedicellatum (9 :3) *gayanus* (6 :6); Association of *S. obtusifolia* + *A. gayanus* (9 :3); Association of *S. obtusifolia* + *P. pedicellatum* (3 :9) Association of *S. obtusifolia* + *P. pedicellatum* (6 :6); Association of *S. obtusifolia* + *P. pedicellatum* (9 :3).

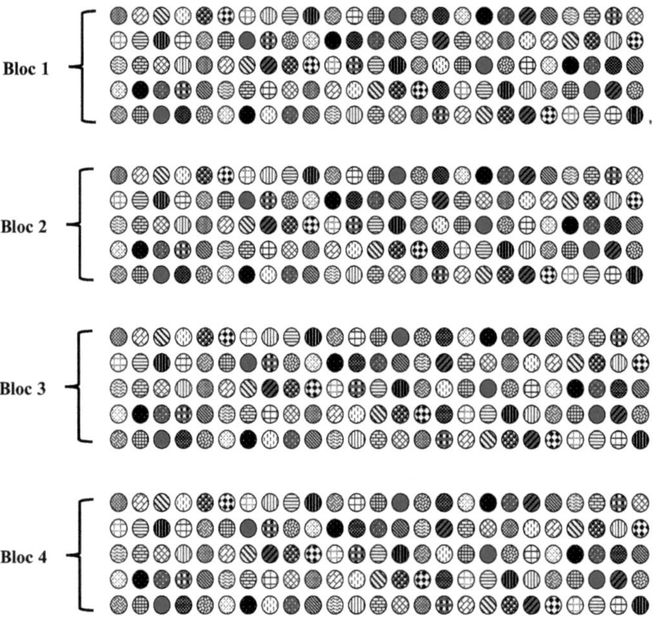

Figure 19Diagram of the greenhouse experimental set-up

Legend: CAC: Field Capacity. Block 1: Control block at 100% CAC; Block 2: Plant stressed at 75% CAC; Block 3: Plant stressed at 50% CAC; Block 4: Plant stressed at 25% CAC. Each circle figure corresponds to either a monoculture species or two associated species with different proportions.

4.4. Competition model applied

Among the methods for studying competition relationships between plants, replacement series experiments have been widely used by several authors (De Wit

,1960; Cousens & O'Neill, 1993; Kumar *et al.*, 2017; Synowiec *et al.*, 2021; Wenda-Piesik *et al.*, 2022) where one species gradually replaces another in a species association at constant overall density (in this study, the constant overall density was 12 individuals). This method enables the process of competition between plants to be understood, in particular its application includes the aspect of inter- and intra-specific interactions between wild plants (Solbrig *et al.*, 1988; Savić *et al.*, 2021). Consequently, competition can be intra-specific when it occurs between individuals of the same species or interspecific when it occurs between individuals of two or more different species (Yang *et al.*, 2019). To this end, this method aimed to elucidate how one species in a species association affects the performance of another, i.e. how *S. obtusifolia* responds in terms of biomass production when competing with A. *gayanus,* C. *mimosoides* and P. *pedicellatum.*

4.5. Determining the quantity of irrigation water

To determine the amount of irrigation water required, the field capacity (FC) of the soil was determined. It corresponds to the mass of water remaining in the soil after rapid drainage over 1 to 2 days. It is expressed as a percentage of the dry weight of the soil. On leaving the autoclave, the substrate was weighed (dry weight) before being watered abundantly. The substrate is then left to stand for 2 days to allow water infiltration, before being weighed again (wet weight). Field capacity is obtained using the following formula:

$$CAC = \frac{X-Y}{Y} \times 100$$

$$CAC = \frac{30,0354kg - 30kg}{30kg} x100 = 0,118$$

The water volume for 100% CAC is : $VE = VS \times 100\% \ CAC$

$$VE = 30L \ x \ 0,118 = 3,54L$$

NB: VE: Volume of water, VS: Volume of soil, CAC: Field capacity, X: weight of wet soil=30.0354 kg, Y: weight of dry soil = 30 kg

4.5.1. Water regime applied

Four (04) blocks, each watered with a different water regime. Water stress was not applied during the germination period, but rather during the growth phase. During the germination phase, the amount of water applied was 2.65 L, corresponding to 75% CAC for all blocks. For the growing phase, corresponding to the application of the water stress regime, plants were watered every other day with a quantity of water determined according to the level of stress. This amount of water or volume of water (VE) varied progressively from one block to the next. CAC is the field capacity and X the portion corresponding to the level of water stress applied. This portion varies progressively from 25%, 50%, 75% and 100%. Thus, 100% CAC is the control regime and the other three are considered as stress regimes depending on the amount of irrigation water. The blocks were under a greenhouse covered with plastic to prevent rainwater infiltration. The application of water stress at the end of germination consisted in watering all plants regularly every two days in the different blocks until the end of the experiment.

4.6. Height and biomass measurement

To characterize the effects of water regime and competition on plants, the height and above-ground biomass of each plant species were determined (Appendix 5). In general, plant biomass and height are used to provide information on the size and aggressiveness of a species relative to its conspecifics, with the aim of determining its competitive capacity (Radosevich, 1987). The height (cm) of each species was measured using a tape measure. Due to the difficulty of accurately separating the roots of individual plants in each pot for analysis, and particularly for perennial grasses, only above-ground biomass was examined (Meekins & McCarthy, 1999; Bolinder *et al.,* 2002). The above-ground dry biomass of the plants was determined after drying in an oven at 65° C for 72 hours, using a precision balance (0.01g). These parameters were used to assess the plants' level of adaptation to water stress (Kagambèga *et al.*, 2019).

4.7. Germination measurement

The number of germinated seeds in the pots was counted every other day for 16 days (appendix 4). A seed is considered germinated when its radicle is visible after the teguments have been pierced.

The germination rate is expressed as the number of germinated seeds in relation to the total number of seeds initially germinated, using the following ratio :

$$\text{Taux de germination (\%)} = \frac{\text{Nombre de graines germées}}{\text{Nombre total de graines}} \times 100$$

The germination rate represents the percentage of germinated seeds at the end of the trial.

The average germination time corresponds to the time at which 50% of the seeds have germinated. The formula used to calculate this parameter is as follows:

$$\text{Temps moy. germ (jour)} = \frac{N1t1 + N2t2 + \cdots + Nntn}{\Sigma Ni}$$

N1=number of seeds germinated between t0 (semi day) and t1 (2 DAS), N2=number of seeds germinated between t1 and t2.

4.8. Sensitivity index

In order to characterize and compare the degrees of tolerance of the different species to water deficit, the stress sensitivity index (SSI) was calculated at the end of the experiment following the different regimes. In this study, the stress sensitivity index developed by Fischer & Maurer (1978) has already been used by other authors (Nana *et al.*, 2009; Kagambèga *et al.*, 2019). This index was determined from the height of the species using the formula below:

$$\text{ISS(\%)} = \frac{H1 - H2}{H1} \times 100$$

With H1, the height values of control plants and H2, the height values of stressed plants. Thus, stress sensitivity indices based on plant height (ISS-H) were determined for each species under the different regimes. The species with the highest sensitivity value is the most sensitive.

4.9 Competition index

The aggressiveness index (A) is a measure of the intensity of competition (Willey & Rao, 1980). It is used in the case of associations between two species (Weigelt & Jolliffe, 2003) and is calculated using the weight of average above-ground dry biomass per species in each pot from the following formulas:

$$RrA = \frac{MA}{PA} \quad \text{an} \quad \text{d} \quad RrB = \frac{MB}{PB} \quad \text{(De Wit \& Van den Bergh, 1965),}$$

Where RrA is the relative biomass of species A, expressing the relationship between species A's average biomass (MA) when grown in association with species B, and its average biomass (PA) when grown in monoculture.

RrB is the relative biomass of species B, expressing the relationship between species B's average biomass (MB) when grown in association with species A and its average biomass (PB) when grown in monoculture.

In this study, A represents *S. obtusifolia* and B each of the other 3 target species. The aggressiveness index can be calculated using Rr (McGilchrist & Trenbath, 1971; Roush & Radosevich, 1985). It is obtained using the following formula:

$$A_A = \tfrac{1}{2} (Rr - Rr)_{AB}$$

$$A_B = \tfrac{1}{2} (Rr - Rr)_{BA}$$

Where, A_A is the aggressiveness index of A and A_B the aggressiveness index of B. An aggressiveness value of zero indicates that the plants in association are all competitive (Connolly *et al.*, 2001). When two species are grown in association in a pot, the species with the higher aggressiveness value is assumed to be the more aggressive and therefore more competitive species, while the less competitive species is the one with a negative aggressiveness value (Meekins & McCarthy, 1999).

4.10. Data analysis

Data obtained from measured parameters (height and biomass) were subjected to a one-factor analysis of variance (Anova) to determine species responses to water regimes. Analysis of variance at the 5% threshold was applied to compare the germination rate of pre-treated and control seeds by species, as well as the average germination time of seeds from the four species. Pearson's correlation test was used to test the correlation between biomass and height of *S. obtusifolia*. Analysis of variance (Anova) at the 5% level revealed a significant difference between the sensitivity indices, so Tukey's Honestly Significant Difference (HSD) test at the 5% level was used to separate the means. The aggressiveness value of each species in mixed cultures was compared with the zero value using Student's t-test (P=0.05). All analyses were performed using R software version 4.1.0 (R Core Team 2021).

5. Results

5.1. Effect of pre-treatment on seed germination rate in the laboratory

Generally speaking, seeds that had undergone pre-treatment had a higher germination rate than those that had not. Pre-treatment of the seeds gave the best germination rates, with a significant difference for *S. obtusifolia* and *C. mimosoides*. Pre-treated and control seeds had germination rates of 95.2% and 31.2% respectively for *S. obtusifolia*, 71.2% and 12% for *C. mimosoides*, 38.4% and 33.6% for *P. pedicellatum* and 34% and 22.4% for *A. gayanus* (Figure 20).

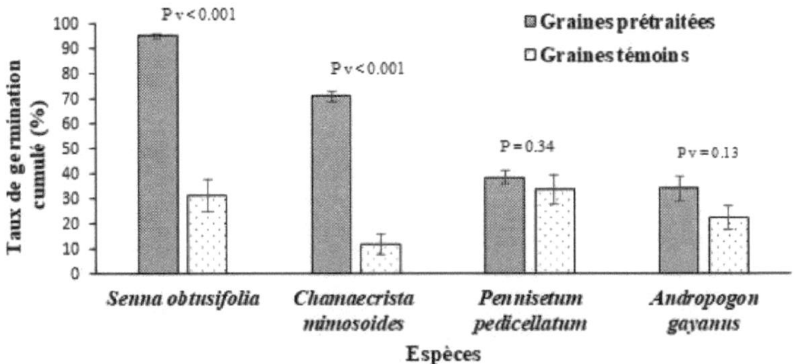

Figure 20Cumulative germination rate (%) of pre-treated and untreated seeds of *Senna obtusifolia, Chamaecrista mimosoides, Andropogon gayanus* and *Pennisetum pedicellatum*.

P-values at the 0.05 level of significance are given directly for comparison of pre-treated and control seeds of each species. Error bars are standard errors.

5.2. Germination kinetics of seeds of four species in the greenhouse

Germination curves show an increase in the germination rate of *S. obtusifolia and C. mimosoides* seeds, reaching maximum rates on 8ème days after sowing (*S. obtusifolia:* 58% and *C. mimosoides:* 68.33%). Germination of *A. gayanus and P. pedicellatum* peaked at 10ème days after sowing, with 34% and 20% germination rates respectively. Germination kinetics curves show three phases: first a latent phase, then exponential acceleration and finally a plateau corresponding to a halt in germination after reaching maximum germination capacity (Figure 21). The results show that during the germination test period, *C. mimosoides* showed the highest germination rates, with a cumulative 68.33%, followed by *S. obtusifolia* (58.66%), *A. gayanus* (34%) and *P. pedicellatum* (20%). In fact, the seeds with the highest germination speed were those pre-treated with 96% concentrated sulfuric acid ($H_2 SO_4$). As for *A. gayanus and P. pedicellatum* seeds, treatment with water for 24 hours showed low germination rates.

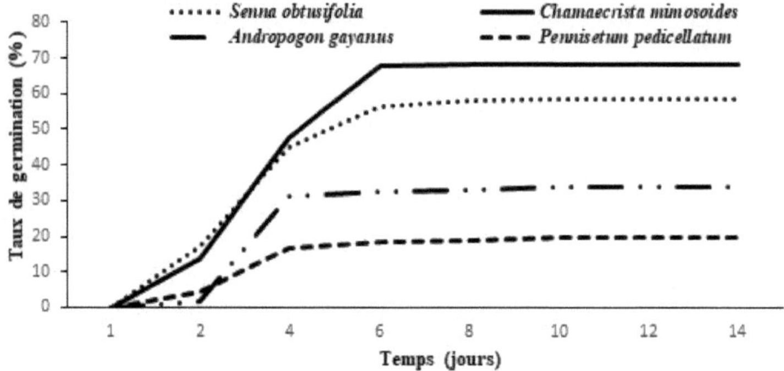

Figure 21Germination kinetics of *Senna obtusifolia, Chamaecrita mimosoides, Andropogon gayanus* and *Pennisetum pedicellatum* seeds.

5.3. Average germination rate and time in the greenhouse

Cumulative germination rates and mean germination time of pre-treated seeds of the species studied during 14 days of testing showed variability according to species. Analysis of variance at the 5% threshold using the Newman-Keuls test showed a highly significant difference (p<0.0001) between the cumulative germination rates of the different species. Comparing the cumulative germination rates of the four (04) herbaceous plants in unstressed water conditions, the results show that *C. mimosoides* has the highest germination rate (68.33%), followed by *S. obtusifolia* (58.66), *A. gayanus* (34%) and *P. pedicellatum* at 20% (Figure 22).

The average germination time, which corresponds to the time at which 50% of the seeds have germinated, did not show a significant difference (p=0.51) according to the analysis of variance at the 5% threshold. This time varies inversely with germination rate. High germination rates were observed with the shortest germination times. Observations showed that the highest average germination time was observed respectively in *C. mimosoides*, which took 4.27 days to see 50% of its seeds germinate. *A. gayanus* takes 4.18 days to have 50% of its seeds germinated and *P. pedicellatum* records 4.20 days for 50% germination. Comparing the average germination times of pre-treated seeds from the four herbaceous species, *S. obtusifolia* is the fastest germinating species in terms of time: 4.01 days for 50% of seeds.

100

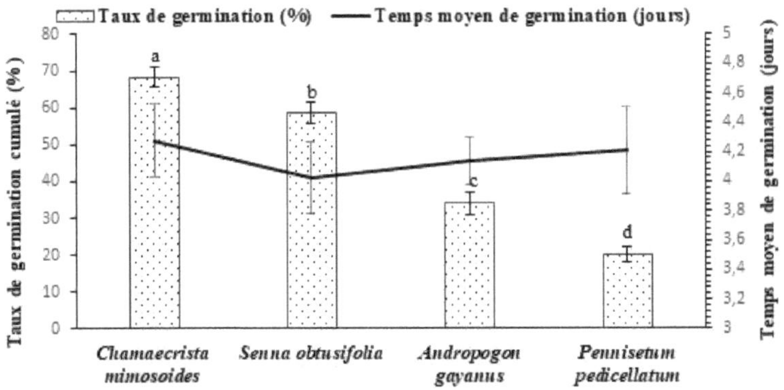

Figure 22Mean time and cumulative germination rate of *Senna obtusifolia, Chamaecrista mimosoides, Andropogon gayanus* and *Pennisetum pedicellatum* seeds.

Values are means, and those with the same letters are not significantly different at the 5% threshold using the Newman-Keuls test. Error bars are standard errors.

5.4. Effects of water deficit on biomass and species height

The water deficit significantly reduced vegetative growth in all four herbaceous species. Indeed, height and biomass values were negatively influenced by stress in the following order: control (100% field capacity, i.e. 100% CAC); very moderate water stress (75% CAC); moderate water stress (50% CAC) and severe water stress (25% CAC) for all species (Figures 23 and 24). Pots in the control block showed a progressive decrease in the size of plants up to the severely water-stressed blocks. A similar trend was observed for above-ground biomass and height (Figures 23 and 24), indicating a positive correlation (r=0.69) between these two parameters.

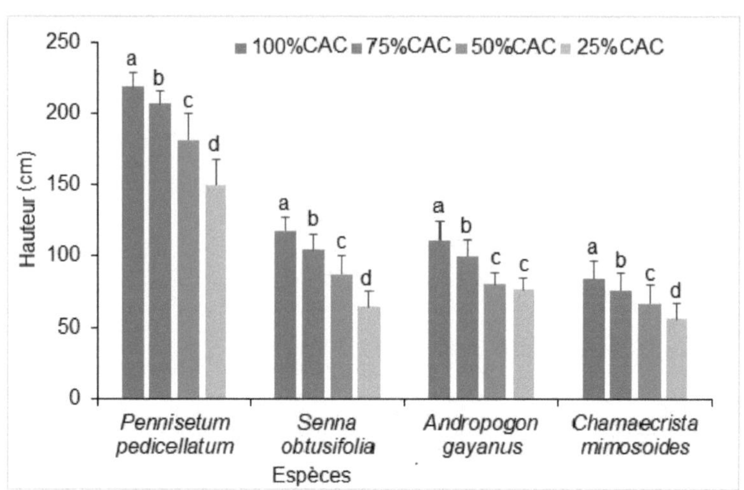

Figure 23Histogram of species height variation as a function of water regime

The letters a. b. c. d on the bar graphs indicate a significant difference (p<0.05).

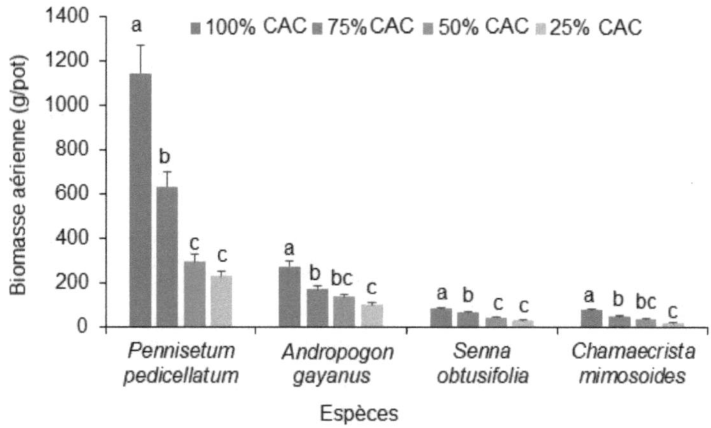

Figure 24Histogram of variation in species biomass as a function of water regime

With reference to plant height, the stress sensitivity index (SSI_H) increased progressively from the 100% CAC water regime to the lowest 25% CAC water regime (Figure 25). Significant variability in the stress sensitivity index (SSI_H) was observed for each species under different regimes (Figure 25). However, the stress sensitivity index did not vary significantly between species in the 75% CAC and 50% CAC regimes. In contrast, in the 25% CAC regime there was a highly significant variation in the SSI-H index (dl = 3; F =26.72 and p < 0.05). With its highest sensitivity index, *S. obtusifolia* was the most sensitive species to water deficit, compared respectively to the other three species: *C. mimosoides*, *P. pedicellatum* and *A. gayanus* (Figure 26).

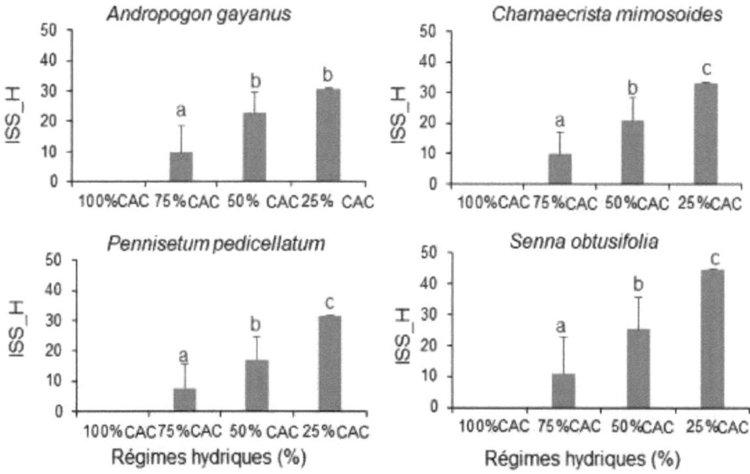

Figure 25 Variations in species sensitivity index to water stress regimes

Letters indicate a significant difference at the 5% threshold according to Tukey's HSD test; error bars represent standard deviation; ISS-H: stress sensitivity index based on plant height.

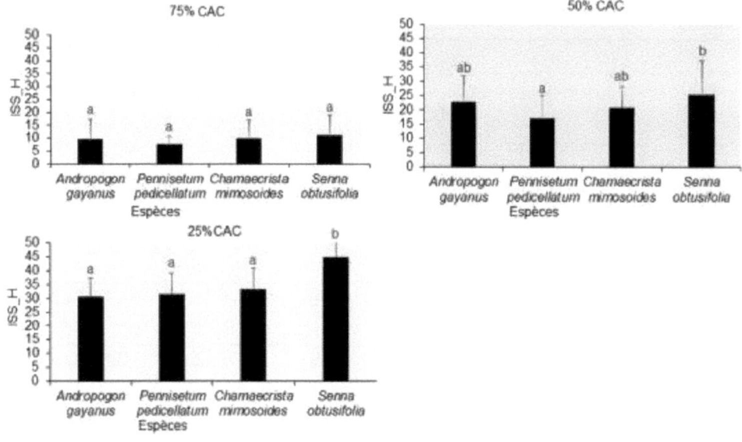

Figure 26Stress sensitivity indices (ISS_H)

The values used are averages and the letters indicate a significant difference at the 5% threshold according to the Tukey HSD test; the bars are standard deviations.

5.5. Effects of competition under water conditions

5.5.1. Biomass production

The above-ground dry biomass of *S. obtusifolia* under all the different water regimes was reduced when *S. obtusifolia* was in association with either *A. gayanus* or *P. pedicellatum* compared with its biomass in monoculture (Table 24). On the other hand, biomass was unaffected in association with *C. mimosoides*. However, for the lowest water regime (25% CAC), the biomass of *S. obtusifolia* in monoculture differed significantly from its biomass in mixed culture with *C. mimosoides*. The biomass of *A. gayanus* and *P. pedicellatum* in monoculture was not significantly different (P> 0.05) from their biomass in mixed culture with *S. obtusifolia* in all water regimes (Table 24). On the other hand, the biomass of *C. mimosoides* in monoculture under severe water stress (25% CAC regimes) was significantly different from its biomass in mixed culture with *S. obtusifolia*.

Table 24Dry above-ground biomass (g) of monoculture and mixed-crop plant species under different water regimes

Water regime	Monoculture		Cultural association			
	Species	Mean± E	Species	Mean ± E	F	P
			SA	59,97 ± 2,10	- 2,77	0,037 *
	S. obtusifolia	90,98 ± 35,97	SC	86,97 ± 28,23	- 0,36	0.984 ns
			S P	60,17 ± 29,41	- 2,75	0,039 *
Indicator	*A. gayanus*	385,57 ± 108,85	AS	376,21 ± 103,88	0,06	0.81ns
	C. mimosoides	93,77 ± 36,71	CS	53,97 ± 28,77	10,92	0,002 ***
	P, pedicellatum	1490,83±688,03	PS	1476,42 ± 68,55	0,003	0.950 ns

	Monoculture		Cultural association			
Water regime	Species	Mean± E	Species	Mean ± E	F	P
Very moderate water deficit	*S. obtusifolia*	74,54 ± 32,25	SA	43,06 ± 21,21	- 3,28	0,001 ***
			SC	68,55 ± 28,98	- 0,62	0.924 ns
			SP	41,66 ± 20,78	- 3,42	0,001 *
	A, gayanus	237,94 ± 93,96	AS	226,39 ± 81,47	0,13	0.72ns
	C, mimosoides	64,71 ± 22,74	CS	32,73 ± 14,91	20,75	0,001 ***
	P, pedicellatum	830,75 ± 377,76	PS	818,33 ± 382,69	0,008	0.93ns
Moderate water deficit	*S, obtusifolia*	47,82 ± 23,34	SA	27,83 ± 15,09	- 3,29	0,001 ***
			SC	43,94 ± 13,72	- 0,64	0.91ns
			SP	24,96 ± 12,03	- 3,77	0,002 ***
	A, gayanus	187,05 ± 66,30	AS	177,85 ± 54,96	0,17	0.68ns
	C, mimosoides	47,97 ± 15,21	CS	25,01 ± 13,43	19,19	0,001 ***
	P, pedicellatum	342,63 ± 130,87	PS	330,98 ± 128,90	0,06	0.80ns
Very severe water deficit	*S, obtusifolia*	38,86 ± 19,92	SA	17,21 ± 5,60	- 4,90	0,001 ***
			SC	22,25 ± 6,03	- 3,75	0,002 ***
			SP	14,66 ± 11,02	- 5,47	0,001 ***
	A, gayanus	147,10 ± 43,18	AS	131,00 ± 23,38	1,61	0,21
	C, mimosoides	25,27 ± 3,33	CS	13,66 ± 3,18	95,14	0,001 *

	Monoculture		Cultural association			
Water regime	Species	Mean± E	Species	Mean ± E	F	P
	P, pedicellatum	290,05 ± 115,70	PS	286,11 ± 115,81	0,009	0.92ns

The values used are averages and (*) indicate significance at the 5% level according to the Tukey HSD test. *** Highly significant; ns not significant. E : Standard deviation

Legend: SA: *Senna obtusifolia* in association with *Andropogon gayanus*; SC: *Senna obtusifolia* in association with *Chamaecrista mimosoides*; SP: *Senna obtusifolia* in association with *Pennisetum pedicelatum*; AS: *Andropogon gayanus* in association with *Senna obtusifolia*; CS: *Chamaecrista mimosoides* in association with *Senna obtusifolia*; PS: *Pennisetum pedicellatum* in association with *Senna obtusifolia*.

5.6. Competition intensity

The value of the aggressiveness index (A) differed significantly between species and proportions of the species combination but not for regime (Table 25). *S. obtusifolia* had a positive aggressiveness value while *C. mimosoides* had a negative aggressiveness in all proportions at almost all water regimes. On the other hand, when *S. obtusifolia* was grown in a mixture with *A. gayanus* or in a mixture with *P. pedicellatum*, the aggressiveness index value of *S. obtusifolia* was negative, while those of *A. gayanus* and *P. pedicellatum* were positive (Table 25). Water regime, species, species proportions and interactions significantly affected aggressiveness (A) but were not affected by their interaction (Table 25).

Table 25Mean value of aggressiveness index (Moy) per plant (A) ± standard deviation (E) for *Senna obtusifolia* and the three target species: *Chamaecrista mimosoides, Andropogon gayanus* and *Pennisetum pedicellatum*.

Water regimes	Proportions	Avg ± E *S.obtusifolia* x *C. mimosoides*	Avg ± E *S.obtusifolia* x *A. gayanus*	Avg ± E *S.obtusifolia* x *P. pedicellatum*
Indicator	75 :25	0,25 ± 0,06*	-0,09 ± 0,02*	-0,12 ± 0,04*
	50 :50	0,1 ± 0,09	-0,21 ± 0,03*	-0,15 ± 0,03*
	25 :75	0,31 ± 0,11*	-0,16 ± 0,09*	-0,25 ± 0,04*
Average		0,22 ± 0,08	-0,15 ± 0,05	-0,17 ± 0,04
Water deficit	75 :25	0,22 ± 0,04*	-0,17 ± 0,04*	-0,19 ± 0,3

very moderate				
	50 :50	$0,17 \pm 0,09^*$	$-0,21 \pm 0,05^*$	$-0,18 \pm 0,3^*$
	25 :75	$0,25 \pm 0,10^*$	$-0,2 \pm 0,04^*$	$-0,29 \pm 0,3$
average		$0,21 \pm 0,07$	$-0,19 \pm 0,04$	$-0,22 \pm 0,3$
Moderate water deficit	75 :25	$0,2 \pm 0,06^*$	$-0,18 \pm 0,03^*$	$-0,2 \pm 0,04$
	50 :50	$0,19 \pm 0,25$	$-0,2 \pm 0,05^*$	$-0,21 \pm 0,03^*$
	25 :75	$0,36 \pm 0,17^*$	$-0,2 \pm 0,03^*$	$-0,23 \pm 0,03$
Average		$0,25 \pm 0,09$	$-0,19 \pm 0,04$	$-0,13 \pm 0,03$
Severe water deficit	75 :25	$0,01 \pm 0,01$	$-0,31 \pm 0,01^*$	$-0,26 \pm 0,03^*$
	50 :50	$-0,01 \pm 0,07$	$-0,16 \pm 0,09^*$	$-0,34 \pm 0,04^*$
	25 :75	$0,18 \pm 0,06^*$	$-0,16 \pm 0,05^*$	$-0,34 \pm 0,05^*$
Average		$0,18 \pm 0,09$	$-0,21 \pm 0,05$	$-0,31 \pm 0,04$

Note. The proportion is the ratio between the percentage of *S. obtusifolia* plants and that of the target species in a pot.

A : T-test is considered significant (*) when it differs ($p \leq 0.05$) from 0

6. Discussion

6.1. Effect of pre-treatment on seed germination of laboratory species

Pre-treatment resulted in a high germination rate of each of the four species in a short space of time. The number of plants of each species required for each pot and for the competition was thus obtained. Pre-treatment of the seeds involved lifting the tegumental inhibition of the pre-treated seeds (Gomgnimbou *et al.,* 2019). *S. obtusifolia* seeds germinated the most, followed by *C. mimosoides*. Pretreatment gave a germination rate of 95.2% for *S. obtusifolia* seeds. These rates are higher than those found by Norsworthy *et al.* (2006) and Thiombiano (2008), which were 78% and 54% respectively. These authors also used sulfuric acid for 5 min to pre-treat *S. obtusifolia* seeds. The difference in germination rates between pre-treated and untreated seeds shows that the seeds are physically dormant, probably due to the hardness of the seed coats. In fact, pre-treatment of the seeds lifted their dormancy and softened the seed coat, allowing rapid water imbibition and air access to the embryo. This hypothesis confirms that of Mbaye (2002).

6.2. Germination and germination rate of seeds of the four species in the greenhouse

The results showed that the germination rate of *S. obtusifolia* was lower than that of *C. mimosoides*. However, the germination rate was still over 50% (58.66%), placing it in the "satisfactory" category (50-60%) according to the catalog of the Centre National de Semences Forestières. In fact, the high germination rate of *C. mimosoides* seeds could be explained by the genetic aspect linked to the intrinsic nature of the species. Even without treatment, the species itself has a high germination rate of 80% under normal conditions, according to Bargali *et al.* (2016). On the other hand, the species' high germination rate could also be due to the effect of pre-treatment. In fact, treatment with sulfuric acid lifted dormancy and also weakened the integument, enabling rapid radicle emergence in all four species. As for germination speed, *Senna obtusifolia* was the species with the highest germination speed: 4.01 days for 50% germination. Even in the wild, *S. obtusifolia* germinates best at the start of the rainy season and reaches the fruiting stage in the middle of the rainy season, giving it an advantage in competition with other native grasses for light, moisture and open space (Dorning *et al.*, 2006). It is also recognized that there are species whose seeds germinate faster than others when subjected to identical living conditions (Thiombiano, 2008).

6.3. Effect of water stress on plant height and above-ground biomass

Height and above-ground biomass of all four species were negatively affected by water stress. This could be explained by the juvenile state of the plants (16-day-old plants) during water deficit induction and the duration of the stress. Indeed, a plant's response to water deficit is complex and depends on the plant's stage of development, the intensity of the deficit, the duration of the stress, the plant's genotype and the state the plant was in at the time of stress (Sawadogo *et al.*, 2006; Aziadekey *et al.*, 2014; Peng *et al.*, 2022). Furthermore, when plants are subjected to long-term severe water stress during the vegetative growth stage, the stress can be severe enough to result in substantial biomass losses (Wijewardana *et al.*, 2019). Plants subjected to soil water deficit experienced a decrease in the performance of their parameters (height and biomass) compared to plants in the control regime. This reduction in growth following a water deficit is considered to be a regulatory mechanism enabling plants to adapt to water restrictions through a reduction in transpiration surface area (Chaves *et al.*, 2002; Yang *et al.*, 2021; Wu *et al.,* 2022).

The sensitivity index (ISS-H) varies significantly for each species under different water regimes. The differences observed between species showing their degree of sensitivity to water stress would be due to the intrinsic characteristics of each species. In fact, grasses are empirically considered "tolerant" to drought (Baruch, 1994; Pardo & Van Buren, 2021). Thus, according to Buldgen (1997), *A. gayanus* is one such grass that is relatively more tolerant and less affected by drought due to a slowdown in the growth of aerial organs in favor of root development.

In the severe stress regime, *S. obtusifolia* with the highest sensitivity index was considered the most sensitive species to water deficit, compared with *C. mimosoides*. *P. pedicellatum* and *A. gayanus* respectively had the lowest sensitivity indices. A plant is tolerant when it is able to maintain its metabolic activity under low water potentials (Kagambèga *et al.,* 2019). From this point of view, species with the lowest water stress sensitivity index, such as *A. gayanus*, *P. pedicellatum* and *C. mimosoides*, are respectively the most tolerant to water deficit. *A. gayanus* and *P. pedicellatum* are plants known for their drought tolerance (Bargali & Bargali, 2016; Issaharou-Matchi *et al.*, 2016). In addition, it is claimed that *C. mimosoides* can withstand low water deficit (Bargali & Bargali, 2016). As for S. *obtusifolia* characterized by the highest sensitivity index, it is the least tolerant to soil water deficit. This shows that *S. obtusifolia* prefers well-drained soils (DAF, 2016). This implies that *S. obtusifolia* absorbs less water in the event of water deficit than other competing species. Our results corroborate those of Moreshet *et al* (1996) who found that when *S. obtusifolia* is in association with *Arachis hypogaea* under water deficit, *Arachis hypogaea* is the one that absorbs more water to the detriment *of S. obtusifolia.*

6.4. Effect of competition

6.4.1. Biomass production

The biomass of *S.* obtusifolia is significantly reduced when grown in association with *A. gayanus* or *P. pedicellatum* compared with its biomass in monoculture. However, the biomass of *S. obtusifolia* is not influenced when grown in association with *C. mimosoides* compared with its biomass in monoculture. On the other hand, the biomass of *A. gayanus* and *P. pedicellatum* is not impacted when grown in association with *S. obtusifolia* compared to their biomass in monoculture. This could be explained by the fact that *A. gayanus* and *P. pedicellatum* have a greater capacity to absorb water and mineral salts through their developed root systems than *S. obtusifolia*. The association of either of these two grasses with *S. obtusifolia* in the same pot can therefore inhibit the normal growth of *S. obtusifolia*. According to Hartvigsen (2000), the growth of certain species is inhibited when they are in association. Furthermore, the combination of *S. obtusifolia* and *C. mimosoides* revealed that in the three least stressed regimes, the biomass of *S. obtusifolia* was unaffected, while that of *C. mimosoides* decreased compared with its biomass in monoculture. However, in the most stressed regime, the water deficit significantly reduced the biomass of both species in association, and in particular the biomass of *S. obtusifolia*. Yet it is recognized that invasive plant species often produce more biomass than their congeners under stress-free conditions. However, according to Kelso *et al.* (2020), some invasive species become less competitive under drought conditions.

6.4.2. Competition intensity

Senna obtusifolia is less competitive than *A. gayanus* and *P. pedicellatum*, with the exception of *C. mimosoides*. In fact, the aggressiveness index of *S. obtusifolia* was negative and lower, while that of *A. gayanus* and *P. pedicellatum* was positive and higher when these species were in association in all proportions and under all water regimes. *A. gayanus* and *P. pedicellatum* had the highest aggressiveness values. This may be due to the efficiency of their photosynthesis mechanism in C_4 and their nitrogen and water use efficiency. According to Zhang *et al.* (2011), a species with the highest competitive ability is generally referred to as a dominant or competitive species, and has a greater capacity to acquire resources and occupy the most space. *S. obtusifolia* is the least competitive species despite its proliferation in Sahelian pastures and also the fact that it is indicated as the dominant species in previous studies (Randell, 1995; Dorning *et al.*, 2006; Gebrekiros and Tessema, 2018). It is often mentioned as the most troublesome or aggressive species when in association with soybean (*Glycine max*), tobacco (*Nicotiana tabacum*), peanut (*Arachis hypogaea*), cotton (*Gossypium hirsutum*) and vegetables (Buchanan *et al.*, 1980; Monks & Oliver, 1988; Webster & Macdonald, 2001). It is also reputed to have many competitive attributes such as rapid growth, deep rooting, rapid biomass accumulation and dense canopy formation to recycle subsoil nutrients, and also its ability to suppress the growth of native species (Hauser *et al.,* 1975; Holt, 1995). However, greenhouse experimentation under water stress revealed that this species is less competitive given its poor performance. This suggests that its invasion is favored more by external factors than by its biology. This result is in agreement with Chambers *et al.* (2014) and Waddell *et al.* (2020) who report that, in general, invasion is facilitated by land use patterns or management activities that disturb native vegetation, and favor the establishment and spread of these invasive species. This disturbance may be partly due to overgrazing (Hobbs & Huenneke, 1992; He *et al.,* 2022).

7. Partial conclusion

The results of the study led to the conclusion that the predictive competitive ability of *S. obtusifolia* is not due to its biology but induced by other extrinsic factors. The first hypothesis, which states that the competitive ability of *S. obtusifolia* is due to its biological characteristics, is invalidated. Moreover, the current invasion of *S. obtusifolia* could be partly due to overgrazing. The current drought in the Sahel is also thought to work against the proliferation of *S. obtusifolia*, which was the species most sensitive to water deficit during the greenhouse experiments. The second hypothesis, which states that water deficit is a factor that improves the performance of *S. obtusifolia* to the detriment of its congeners, is invalidated. This experiment on competition in the greenhouse gives an idea of the performance of *S. obtusifolia* in the wild and the factors behind its invasion. Furthermore, sustainable management of Sahelian pastures would help reduce the invasion of *S. obtusifolia* and provide more fodder for livestock. Would a less disturbed environment be favorable to the proliferation of *S. obtusifolia*?

CHAPTER V: INTERANNUAL MONITORING OF THE EFFECT OF FENCING ON THE FLORISTIC DYNAMICS OF AREAS INVADED BY *SENNA OBTUSIFOLIA* (L.) H. S. IRWIN & BARNEBY IN THE SAHELIAN ZONE

1. Introduction

In the Sahelian zone, where livestock farming is the main economic activity (Ayantunde *et al.*, 2020), we are unfortunately witnessing a continuous degradation of natural pastures, which are the basis for feeding extensive livestock (Aboh *et al.,* 2012). This degradation is due to a combination of factors, including climate change and human activities, which play a predominant role. Today, areas set aside for livestock farming suffer from a lack of effective management and are therefore increasingly degraded. This degradation has led to a reduction in forage, forcing livestock to overgraze. As a result, pastures can no longer regenerate properly, and increasingly degraded rangelands have led to the appearance of unpalatable and invasive species (Kiéma *et al.*, 2007). *S. obtusifolia* is one of these herbaceous species of low fodder value (Sawadogo, 2011), the most widespread in Sahelian pastures. Its invasion therefore leads to a reduction in fodder quality and a consequent shortage of fodder. However, field observations show that less degraded areas are less invaded by *S. obtusifolia*. These areas tend to contain a high diversity of flora. This is the case of the protected areas found in some Sahelian localities. Unfortunately, these areas are becoming increasingly rare in this Sahelian zone, due to strong anthropic pressure, particularly from overgrazing. It is therefore necessary to promote these protected areas in an effective and sustainable way. With this in mind, numerous natural resource management actions have been initiated and are supported by development partners with the aim of improving and restoring natural resources (Kiéma *et al.,* 2012).

Through this activity, the following objectives were formulated:

- To evaluate the effect of protection duration on the proliferation *of S. obtusifolia* ;
- Determine the effect of protection duration on the restoration of herbaceous vegetation.

The hypotheses that support these objectives are :

- The duration of protection does not favour the proliferation of *S. obtusifolia* ;
- The duration of the protection helps to restore the vegetation.

2. Methodology

2.1. Study area

The study was carried out in the Sahelian zone of Burkina Faso, specifically in the communal forest of the commune of Dori, capital of the Séno province, 265 km from Ouagadoudou (Figure 27). The communal forest was created in 2017 on a site formerly used for grazing, fields and dwellings. The forest site was fenced off with stakes and wire fencing when a portion was invaded by *S. obtusifolia.* The soil restoration techniques used on this site were mainly half-moons combined with scarification. This forest benefits from integral protection, covering an area of 207.4 ha, of which 132.42 ha is reserved for long-term forage production (IUCN, 2017). The main villages surrounding the forest are Bafélé (Dori sector 8), Djomga. The vegetation of this forest is characterized by shrub steppe for 21.6% of the forest area and grassy steppe for 78.4% of the forest area. The soils of the communal forest are hydromorphic, iso-humic, gravelly, sodic, iron sesquioxide types (PCD, 2015). The commune's production systems are characterized by the coexistence of agropastoral activities (IUCN, 2017). The Peulh are the majority indigenous socio-cultural group (76.6%). The other socio-cultural groups are the Bella (14.66%), Rimaibé (2%), Sonraï (2%), Djerma (0.66%), Gourmatché, Mossi, Dagara, Gourounsi and Bissa (MEF, 2008).

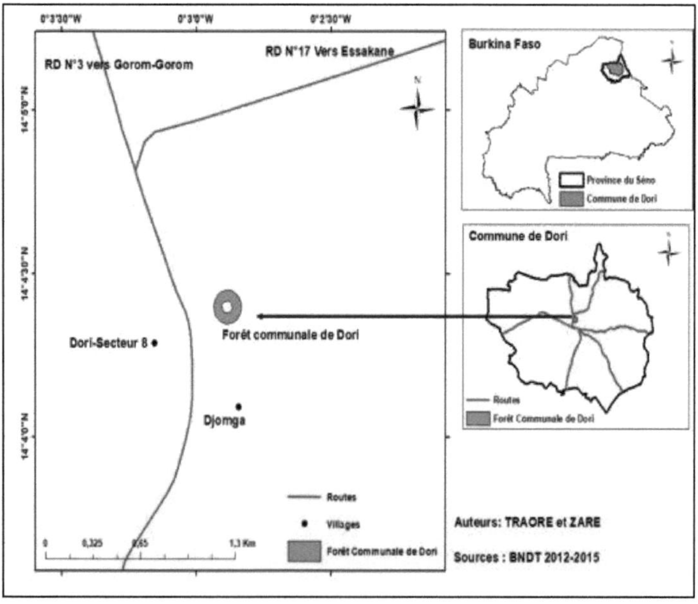

Figure 27Dori communal forest location map

2.2 Data collection

Annual monitoring of changes in the density and biomass (weight) of S. *obtusifolia* and the biomass of other herbaceous species was carried out for four consecutive years from 2017 to 2020 in the Dori communal forest. The experimental set-up used to monitor the dynamics of S. *obtusifolia* consisted of 40 permanent plots, each with a surface area of 100 m². These plots were placed in areas of the forest invaded by S. *obtusifolia.* Each plot was marked with four white-painted iron stakes, and the GPS coordinates of each plot were noted to facilitate identification. Invasion levels are those mentioned in chapter III.

2.3 Floristic inventory and biomass harvesting

The floristic inventory was carried out in 40 permanent plots, where an exhaustive list of herbaceous species was drawn up, with each species assigned an abundance-dominance coefficient according to the Braun-Blanquet (1932) scale. Monitoring was carried out over four years, from 2017 to 2020. Within each plot, three (03) sub-plots or plots of 1 m² were randomly installed to estimate herbaceous density, height and biomass, followed by a biomass cut. The average height and total number of individuals (density) of S. *obtusifolia* were noted beforehand. Subsequently, all the grass in the subplot is mown flush with the ground (Alhassane *et al.*, 2018; Moussa *et al.,* 2018) using a sickle and placed on a plastic sheet. By sorting, S. *obtusifolia* is separated from the other grasses and the fresh weight of each of the piles is determined using an electronic spring scale, precision 50 kg ± 10 g. Using a 5 kg ± 1 g precision electronic balance, a 100 g sample was taken from each pile. These herbaceous samples were pre-dried on site and then oven-dried in the laboratory for 72 hours at 75°C to obtain a constant dry weight, and then used to determine dry biomass (Ouédraogo, 2009; Zerbo, 2011). The pastoral value of permanent plots is estimated as follows:

Forage species categories, i.e. herbaceous species with no pastoral value (SVP), species with low pastoral value (FVP), species with medium pastoral value (MVP) and species with good pastoral value (BVP), were defined on the basis of the 4-class specific quality index scale (Ngom *et al.,* 2012; Sawadogo, 2011) (Table 26). For species for which the authors use the 6-class specific quality index scale and for woody regeneration, correspondences were established in order to identify forage species categories (Table 27). The recovery rates of the different forage species categories were used to construct forage spectra.

Table 26Classification of forage species categories according to the four-class specific quality index scale

Specific 4-class quality index scale	0	1	2	3
Forage species categories	PLE ASE	FVP	MVP	BVP

Table 27Classification of forage species categories according to the four- and six-class specific quality index scale

Specific 6-class quality index scale	0	1	2	3	4	5
Palatability qualifier for woody regeneration	NA	PA		A		TA
Forage species categories	PL EA SE	FVP		MVP		BVP

2.4. Data analysis

The floristic diversity of the herbaceous vegetation at these study sites is assessed by listing plant species, genera, families, biological types and phytogeographical types based on the Burkina Faso catalog of vascular plants.

Analyses of variance (ANOVA) at the 5% threshold with R software version 4.1.0 were used to test the influence of protection duration on *S. obtusifolia* parameters. The Shapiro-Wilk normality test with R software version 4.1.2 revealed that data related to vegetation taxon richness were not normally distributed. Consequently, the non-parametric Kruskal-Wallis test was used for this purpose. Pc-ord 6 software was used to calculate species richness, Shannon diversity and Piélou equitability. Species richness (SW) indicates the number of species inventoried in a well-defined area (plot) . Shannon's diversity index (H) quantifies the heterogeneity of an environment's specific diversity. H generally varies on average from 0 to 5 bits. Its high values reflect species abundance, and therefore favorable environmental conditions. Piélou's equitability (E) represents the equi-repartition of all individuals present per species. It varies from 0 to 1 and indicates a regular distribution of individuals between species with values close to 1, and the dominance of one species or a small number of species with values close to 0. The method for analyzing data on the pastoral value of plots is identical to that used in Chapter II.

3. Results

3.1. Effects of protection duration on density, height and biomass *of Senna obtusifolia*

The duration of protection had a significant effect ($p < 0.05$) on the density, height and biomass of *S. obtusifolia*, as well as on the biomass of other grasses. In general, the density, height and biomass of *Senna obtusifolia* decreased with the duration of protection, whereas the biomass of other grasses increased significantly (Table 28). In 2018, the drop in rainfall contributed to a regression in the density, height and biomass of *S. obtusifolia* and an increase in the biomass of other grasses. The increase in the standardized rainfall index in 2020 favored an increase in the biomass of other grasses and a decrease in the density, height and biomass of *S. obtusifolia* (Table 28).

Table 28 Variation in density, height, biomass of *Senna obtusifolia* and biomass of other herbaceous plants as a function of monitoring years and rainfall.

S. obtusifolia Year	ISP	Density (ind./ha)	Height (cm)	Biomass (t/ha)	Biomass Other (t/ha)
	-	Mean ± sd	Mean ± sd	Mean ± sd	Mean ± sd
2017	0,25	80,23 ± 70,98.10^4[a]	103 ± 24,67[a]	2,30 ± 1,61[a]	1,17± 0,69[a]
2018	-0,15	43,13 ± 34,07.104[b]	84,61 ± 20,36[b]	1,25 ± 0,86[b]	2,0 ± 0,93[b]
2019	0,75	25,70. ±25,08.104[c]	76,08 ± 21,15[c]	0,704 ±5,2[c]	2,66 ± 1,11[c]
2020	3	12,72 ± 6,51. 10^4[c]	63,67 ±18,62[d]	0,46 ± 0,36[c]	3,46 ± 1,3 1[d]
F	-	44,66	54,02	64,44	79,36
P	-	0,0001	0,0001	0,0001	0,0001

ISP : Indice Standardisé des Précipitations ; Ind/ha : individuals per hectare ; cm ; centimètre ; t/ha : tonnes per hectare ; Mean ± sd :

3.2. Interannual dynamics of the taxonomic flora of herbaceous vegetation

Over the four years of monitoring, a total of 128 species in 27 families and 80 genera (Table 28, Appendix 7) were recorded. In general, the flora is dominated by the

Poaceae (40%), followed by the Fabaceae (22%). Rubiaceae account for 12% and Cyperaceae for 10% of the species inventoried (Figure 29).

Species richness increases with the duration of protection. The number of species, families and genera has increased significantly from 2017 to 2020. From 67 species in 2017 to 100 species in 2020 (Table 28).

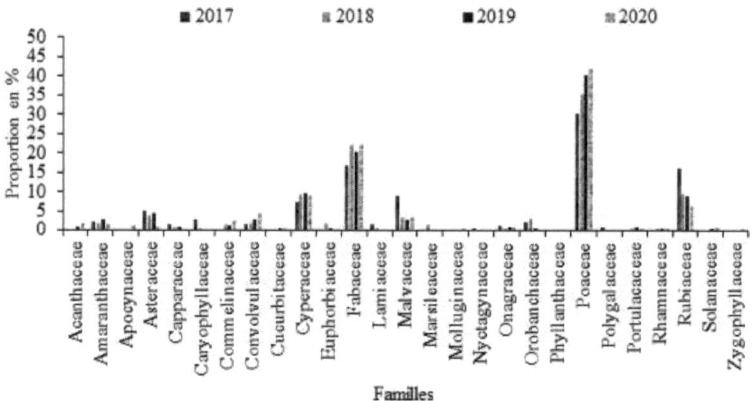

Figure 28Histogram of flora families by year

Table 29Taxonomic richness of herbaceous vegetation

Follow-up years	Number of species	Type	Family
Global	128	80	27
2017	67	47	18
2018	82	55	21
2019	96	63	24
2020	100	60	18

3.3.Interannual dynamics of biological and phytogeographical types of herbaceous vegetation

The biological representation of the species recorded shows an average predominance of therophytes (25 species) compared with other biological types (Figure 29). As for phytogeographical types, species of Pantropical and Afrotropical origin (Figure 30) are in the majority. The least represented are Afro-Malagasy and Cosmopolitan (Figure 30).

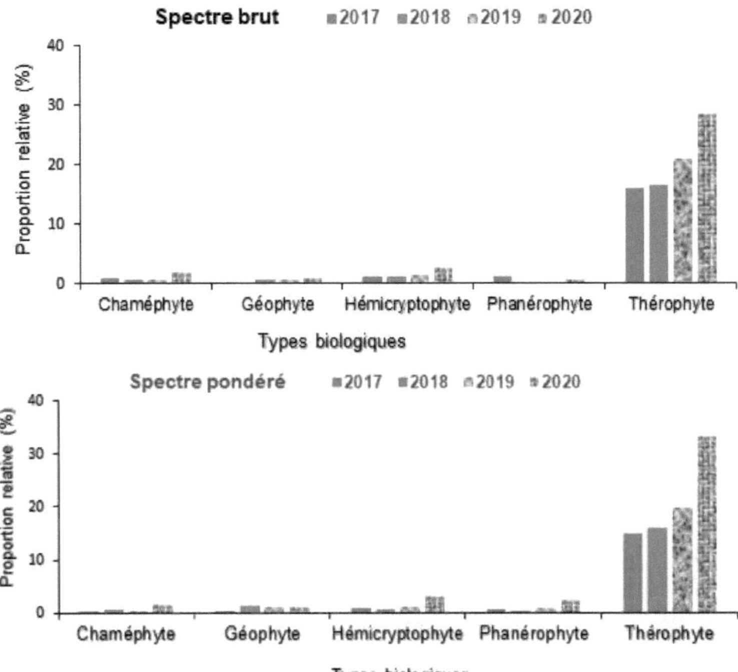

Figure 29Biological types (2017 to 2020)

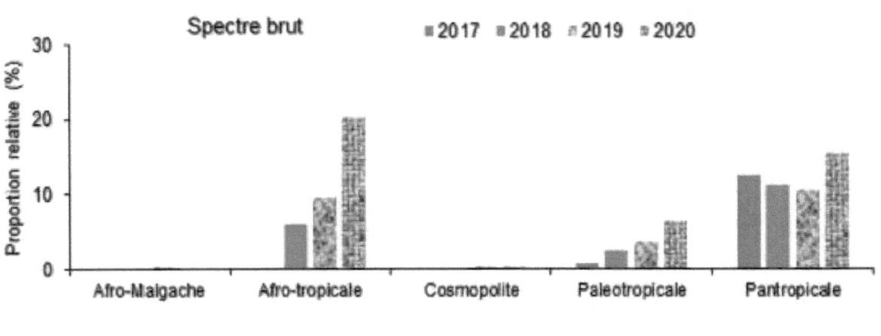

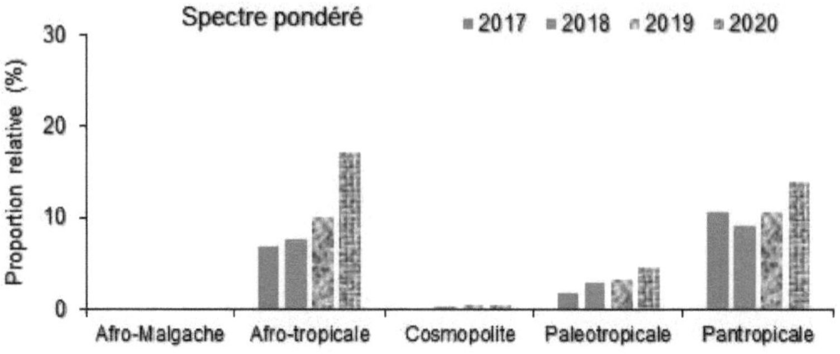

Figure 30Phytogeographic type (2017 to 2020)

3.4. Floristic diversity and similarity of herbaceous vegetation according to monitoring years

Mean species richness per plot, Shannon diversity indices, Piélou fairness and Simpson index varied significantly from year to year (P <0.05). The first three years (2017; 2018 and 2019) were the least diverse (Table 30), in contrast to the fourth year (2020), which recorded an increasing number of species (15.32 ± 4.63 species). The Shannon diversity index varies in the same direction as species richness, with an average value of 1.58 ± 0.19 for the last three years, compared with 2.27 ± 0.34 for 2020. In 2017, the Piélou and Simpson indices were respectively the lowest (0.7 ± 0.160; 64 ± 0.18), but in 2020, the value of these indices has increased considerably (0.84 ± 0.06; 0.86 ± 0.07) (Table 30).

The composition of herbaceous vegetation varies from one year to the next and is mainly influenced by the duration of protection. Sørensen's similarity index reveals no similarity in vegetation between the four years of monitoring (Table 31).

Table 30Alpha diversity index (mean ± standard deviation)

Years	S	H	E	D
	M±sd	M±sd	M±sd	M±sd
2017	7,85 ±3,85[b]	1,39 ± 0,49[b]	0,70 ± 0,16[b]	0,64 ± 0,18[c]
2018	8,17± 3,54[b]	1,56 ± 0,41[b]	0,78 ± 0,11[a]	0,73 ± 0,11[b]
2019	10,27 ± 3,81[b]	1,78 ± 0,31[b]	0,79 ± 0,05[a]	0,78 ± 0,07[b]
2020	15,32 ± 4,63[a]	2,27 ± 0,34[a]	0,84 ± 0,06[a]	0,86 ± 0,07[a]
F	30,11	36,86	12,27	24,85
P	< 0,001	< 0,001	< 0,001	< 0,001

Levels not connected by the same letter are statistically different. S: species richness index; H: Shannon diversity index; E: Piélou equitability index; D: Simpson diversity index; F: Fisher's F-value; P: p-value or probability at the 0.05 significance level.

Table 31Sørensen Similarity Index comparing by year of follow-up

Year	2017	2018	2019	2020
2017	1			
2018	0,3973	1		
2019	0,2279	0,4361	1	
2020	0,1597	0,267	0,4559	1

3.5. Dominant herbaceous species over four years of monitoring

During the four years of monitoring, a succession of herbaceous species of different invasion levels was recorded, with different levels of dominance. In 2017, the dominant species were *Senna obtusifolia (L.) H. S. Irwin & Barneby*, *Schoenefeldia gracilis* kunth., *Spermacoce filifolia* J.-P. Lebrun et stork , *Acanthospermum hispidum DC*, *Spermacoce stachydea DC*, *Panicum laetum* kunth, *Chamaecrista mimosoides* (L.) Greene and *Chloris pilosa* schum. Most of these species have little or no pastoral value (Table 32).

In 2018, in addition to *Senna obtusifolia*, other species also proved dominant. These were: *Pennisetum pedicellatum* Trin., *Panicum laetum J.-P. Lebrun et stork* , *Brachiaria lata* (Schumach.) C. E.Hubbard , *Zornia glauchidiata* C.Rchb. ex DC., *Spermacoce filifolia* J.-P. Lebrun et stork , *Schoenefeldia gracilis* kunth, *Echinochloa colona* (L.) Link , *Chamaecrista mimosoides* (L.) Greene and *Chloris pilosa schum.*

In 2019, the third year of protection, *Senna obtusifolia* is still the most dominant species, followed by a number of species with pastoral value: *Pennisetum pedicellatum* Trin., *Spermacoce filifolia* J.-P. Lebrun et stork , *Chamaecrista mimosoides (L.) Greene.*, *Brachiaria lata* (Schumach.) C. E.Hubbard., *Panicum laetum J.-P. Lebrun et stork ..*, *Schoenefeldia gracilis* Kunth, *Zornia glauchidiata C.Rchb. ex DC*, *Echinochloa colona* (L.) Link, *Chrysanthellum Americanum (L.) Vatke, Setaria barbata (Lam.) Kunth, Dactyloctenium aegyptium* (Linn) Wild. and *Cyperus rotundus L.*

For the last year of monitoring (year 2020), the dominant species are *Schoenefeldia gracilis* kunth, *Brachiaria lata* (Schumach.) C. E.*Hubbard, Pennisetum pedicellatum* Trin.., *Alizicarpus ovalifolius* (S. et Th.) Leon , *Spermacoce filifolia J.-P. Lebrun et stork , Panicum laetum J.-P. Lebrun et stork , Chamaecrista mimosoides (L.) Greene, Zornia glauchidiata C.Rchb. ex DC and Cyperus iria* Linn(Table 32).

For the first three consecutive years of 2017, 2018 and 2019, *Senna obtusifolia* was the dominant species, with decreasing relative importance values of 44.59, 33.05 and 13.47 respectively. In 2020, on the other hand, *Schoenefeldia gracilis, Pennisetum pedicellatum* and *Alizicarpus ovalifolius* are the most dominant species, with relative importance values greater than ten. The relative importance of *Senna obtusifolia* decreased to a value of 3.62. In the study site, as protection lasts, certain species appear. These include Andropogon spp. in herbaceous species and *Vachellia* spp. in woody regeneration (Table 33).

Table 32Herbaceous species of relative importance (IR ≥ 5)

Species	Vp	2017	2018	2019	2020
Acanthospermum hispidum	*PLEA SE*	6,48	2,51	1,11	0,19
Alysicarpus ovalifolius	*BVP*	1,47	2,96	2,86	7,27
Brachiaria lata	**BVP**	**3,93**	**7,58**	**9,03**	**12,27**
Chamaecrista mimosoides	*MVP*	5,26	6,09	9,30	5,80
Chloris pilosa	*BVP*	5,04	5,87	4,64	4,02
Chrysanthellum americanum	*PLEA SE*	0,37	3,49	6,21	1,73
Cyperus iria	*MVP*	4,83	3,19	4,44	5,05
Cyperus rotundus	*MVP*	1,11	4,85	5,29	3,51
Dactyloctenium aegyptium	*BVP*	4,21	3,68	5,58	4,85
Echinochloa colona	*MVP*	4,69	6,33	6,47	2,99
Panicum laetum	*MVP*	5,26	8,05	8,79	6,16
Pennisetum pedicellatum	**BVP**	**2,17**	**10,97**	**11,54**	**11,69**
Schoenefeldia gracilis	**MVP**	**7,70**	**6,49**	**8,27**	**15,28**
Senna obtusifolia	**FVP**	**44,59**	**33,05**	**13,47**	**3,62**
Setaria barbata	*BVP*	1,26	1,01	5,67	1,03
Spermacoce filifolia	*PLEA SE*	6,89	7,07	11,26	6,84

Species	Vp	2017	2018	2019	2020
Spermacoce stachydea	*PLEA SE*	6,31	3,73	3,68	1,40
Zornia glochidiata	*BVP*	2,51	7,38	6,67	5,46

A species is dominant if its relative importance index is greater than or equal to 5.

Vp: Pastoral value; MVP: Average pastoral value; BVP: Good pastoral value; FVP: Low pastoral value; SVP: No pastoral value

Table 33Herbaceous and woody species that appeared with the duration of protection

Type	Plant species	2017	2018	2019	2020
Herbaceous	*Aeschynomene indica*	-	-	0,57	1,03
	Andropogon fastigiatus	-	-	0,57	2,32
	Andropogon gayanus	-	-	-	1,11
	Andropogon pseudapricus	-	-	0,56	1,00
	Hackelochloa granularis	-	-	-	1,86
	Indigofera stenophylla	-	-	0,77	0,51
	Ipomoea triloba	-	-		0,76
	Ipomoea asarifolia	-	-	-	0,73
	Ipomoea eriocarpa	-	-	-	1,62
	Stylosanthes erecta	-	-	-	3,70
	Tribulus terrestris	-	-	-	0,73
Woody regeneration	*Dichrostachys cinerea*	-	0,36	0,29	0,51
	Prosopis juliflora	-	-	-	1,13
	Vachellia nilotica	-	-	0,28	0,51
	Vachellia seyal	-	-	0,29	0,89
	Vachellia sieberiana	-	-		0,70
	Vachellia tortilis	-	-	1,91	3,34

3.6. Interannual dynamics of pastoral value as a function of invasion levels

The graphical representation of the forage shows that the cover rates of species of Low Pastoral Value (LPV) and No Pastoral Value increase with the invasion of *S. obtusifolia* in the first two years of monitoring. However, the cover rate of species of Medium Pastoral Value (MVP) and Good Pastoral Value decreased (Figure 31). However, in the last two years of monitoring (2018 and 2019), the cover rates of species of Low Pastoral Value (FVP) and no Pastoral Value have dropped considerably in favor of species of Medium Pastoral Value (MVP) and Good Pastoral Value (BVP) (Figure 31).

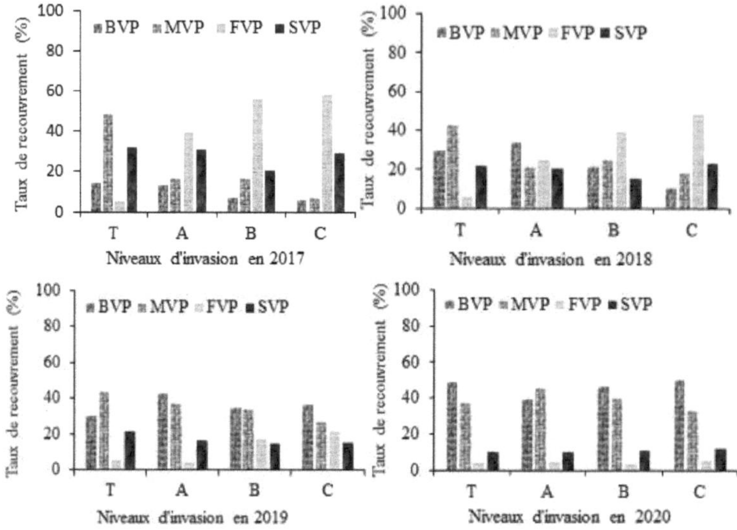

Figure 31 Graphical representation of forage from 2017 to 2020

BVP: Good pastoral value; MVP: Average pastoral value; FVP: Low pastoral value; SVP: No pastoral value

4. Discussion

4.1. Effects of protection duration on biomass, height and density of *Senna obtusifolia*

The length of time the Dori communal area has been protected has led to a reduction in the proliferation of *Senna obtusifolia* and an increase in the density of other herbaceous species. This decline is characterized by a decrease in biomass, density and height of *S. obtusifolia*. These results could be partly explained by the cessation of grazing in this protected area. In fact, thanks to the fence erected and the use of restoration techniques such as half-moons combined with scarification, this area, which is now inaccessible to livestock, has favoured the restoration of several species, including palatable species which compete with less palatable species such *as S. obtusifolia* for space. In addition, the spread of dry *S. obtusifolia* seeds by cattle (Thiombiano *et al.*, 2012) has been interrupted by the protection of the site. This resulted in the non-renewal of the *Senna obtusifolia* seed bank in the soil, leading to a reduction in *Senna obtusifolia* density over time and an increase in the number of other herbaceous species. More specifically, the duration of protection in the last two years has led to the regression of *Senna obtusifolia* and species with low/no pastoral value. The regression of these species favoured the release of space in favour of species with high forage values at all invasion levels. Our results are in line with those of Konedera *et al* (2010), who found that during a 6-year follow-up, *Senna obtusifolia* and *schoenefeldia gracilis* were among the most dominant species, to be replaced by other high forage value species such as *Brachiaria lata*, *Penicetum pedicellatum*, *Zornia glauchidiata*, etc.

The drop in *S. obtusifolia* density after four years may be due to the fact that the longer the seed remains in the soil, the lower its viability. Indeed, some authors claim that *S. obtusifolia* seeds have hard envelopes and can remain viable in the soil seed bank for more than 5 years (Creel *et al.,* 1968; Egley and Chandler 1983; Senseman and Oliver, 1993). However, other authors (Ewart, 1908; Doll *et al.*, 1976; Egley *and* Chandler, 1978; Binggeli, 2005;) have found that in dry storage, *S. obtusifolia* seeds lose their viability quite rapidly. These authors showed that seeds stored for three years had an overall germination of 22%. They also stated that *S. obtusifolia* seeds stored for nine years had a germination of between 9% and 10% of seeds buried in the soil.

The control levels were not infested over time by *Senna obtusifolia*, in contrast to the more heavily infested invasion levels, where *S. obtusifolia* density, biomass and height declined considerably over time. These results indicate the role that animals can play in the spread of *Senna obtusifolia* seeds within invaded pastures and the pressure on other species. Their absence led to a decline in the proliferation of *Senna obtusifolia* within the protected area.

4.2. Effects of fencing on the dynamics of herbaceous vegetation

4.2.1. Taxonomic richness of herbaceous vegetation

In the study sites, and during the four years of monitoring, we observed a dominance of the Poaceae and Fabaceae families over the other families. This dominance of Poaceae is due to the fact that grasses are the most represented species in the savannah (Beker *and* Muler, 2007; Zizka *et al.*, 2015). As for Fabaceae, which occupy second place, this could be explained by the fact that *S. obtusifolia*, being a fabaceae, imposes its physiognomy (an environment dominated by the *S. obtusifolia* population) on the herbaceous vegetation inventoried during this study through its high cover rate. Specific richness also increased with the duration of protection. In fact, the cessation of grazing favored the reestablishment of several species, including the palatable ones. As for the results relating to biological types, the high representation of therophytes is linked to the degree of anthropic disturbance that helped free up space for the establishment of annual species. Our results corroborate those of Goudard (2007). The most representative phytogeographical types, being of pantropical and tropical African origin, are certainly those found most frequently in West Africa.

4.2.2. Specific diversity and similarity of herbaceous vegetation according to monitoring years

Specific diversity varies from year to year. This variation is mainly influenced by the duration of protection. In 2017, the Shannon and Simpson diversity index values were 1.39 bit and 0.64 respectively, and the Piélou equitability value was 0.70. Aboh (2008) found that the uneven distribution of species was linked to the dominance of invasive plants. Thus, the results obtained reflect the low diversity of species and their uneven distribution due to the dominance of *S. obtusifolia* in the study site. From 2017 to 2020, the Shannon and Simpson diversity index values increased significantly. The same applies to Piélou's equitability. The gradual increase in the values of these indices reflects a clear improvement in species diversity and their equal distribution in an environment that is increasingly favorable to their establishment, due to the regression of *S. obtusifolia* in the study site.

Sørensen's similarity index reveals no similarity in vegetation during the four years of monitoring after the protection. This could be explained by the natural process of plant succession, especially favored by protection. In the course of this succession, certain species disappear to the benefit of others. According to Abourouh *et al* (2005), the elimination of grazing encourages a revitalization of the vegetation and allows certain plant species to regenerate. Métro and Sauvage (1955) add that the intensity of this natural regeneration is proportional to the duration of protection.

4.2.3. Effect of pedology (soil enrichment) on the regressive dynamics of *Senna obtusifolia*

Senna obtusifolia is a species whose proliferation is also favoured by fertile soils (Tungate *et al.*, 2002). However, if despite the progressive fertilization of the protected environment, there is a regression in the dynamics of *S. obtusifolia,* this could be explained by the fact that fertilization of the environment alone is not sufficient for the effective establishment of the species. Soil fertility must be accompanied by environmental disturbance. According to Dunlop *et al.* (2006); Solomon and Yayneshet (2014), disturbed environments are more favorable to the proliferation of *S. obtusifolia* than undisturbed environments.

4.2.4. Dominant herbaceous species over four years of monitoring

The predominance of *S. obtusifolia* over other herbaceous species during the three consecutive years of monitoring would appear to be due to the high stock of seeds of this species disseminated during grazing, and their good resistance and germination capacity. Also, soil restoration practices using half-moons combined with scarification from the first year of protection contributed to a proliferation of *S. obtusifolia* in the communal forest. These observations are similar to those of Kiéma *et al* (2012), who found that soil enrichment by half-moon practices favored an increase in legumes, particularly *S. obtusifolia*. However, the sharp decline in the species in the last year of monitoring is thought to be due to the depletion of seed stocks not renewed by the suppression of grazing, and to the occupation of space by other species that were prevented by grazing. *Pennisetum pedicellatum, Alizicarpus ovalifolius* as well as andropogonaceae and Sahelian woody regenerations are species that are highly palatable to animals. The number of years required for an evolutionary change in the number of plant species following a change in practices varies greatly from one context to another, and often depends on a combination of factors. Jeangros and Berthola (2002) observed the appearance of ten new species in four years in a protected area.

5. Partial conclusion

Protected areas have contributed to a reduction in anthropogenic disturbance, thus reducing the proliferation of *S. obtusifolia* . The first hypothesis, which states that the duration of protection does not favour the proliferation *of S. obtusifolia,* is confirmed. This protection also favoured the restoration of plant species with good forage value. The second hypothesis, which states that the duration of protection helps to restore vegetation, is confirmed. The protection did, however, encourage the appearance of several other herbaceous and woody species that had disappeared due to human disturbance. These results could lead to a biological perspective of reducing the proliferation of the species in favor of species of good forage quality. Defensing could be an alternative for restoring the productivity and forage value of degraded pastoral areas. This practice could be popularized in the country's pastoral zones.

GENERAL CONCLUSION, OUTLOOK AND RECOMMENDATIONS

General conclusion

Senna obtusifolia is a species that has become invasive through human and animal contact. Human agropastoral activities have directly or indirectly contributed enormously to the spread of *S. obtusifolia* seeds, either by mowing dry stems to encourage seed fall, or by transporting these stems. As for the animal, forced to consume the seeds of the species in the dry season due to the scarcity of fodder, it has probably become the greatest vector of dissemination of the species in the Sahel. This work focused on the factors that drive the invasion of *S. obtusifolia*. The competitive advantages of *S. obtusifolia* are due to external factors, all the more so as this native species used to be located mainly in lowlands. In addition, its vulnerability was highlighted by the fact that it was the most sensitive to water deficit and less competitive with *A. gayanus* and also *P. pedicellatum*. These two native grasses are highly appreciated by ruminants, but their density has been greatly reduced in pastures by overgrazing. However, the traits of *S. obtusifolia*, characterized by a high germination rate, rapid growth and low palatability, give it a certain competitive advantage over other grasses and enable it to colonize space. Furthermore, the elimination of factors favoring its proliferation shows the species' inability to impose itself on other species over a prolonged period in an undisturbed environment. The use of fencing is therefore an effective means of controlling the species' expansion. Moreover, *S. obtusifolia* is not only an undesirable species, it is also a source of food, particularly during the lean season, and an economically profitable craft product for local populations, especially in the Sahel. Enhancing the value of this species would be more profitable and could also be a means of controlling its invasion.

Outlook

This study highlighted the problem of managing invasive plants in degraded pastures in relation to ecological and social factors. It revealed the factors behind the invasion of *Senna obtusifolia* and its impact on ecosystem services. It also highlighted management alternatives for this species. It would make a significant contribution to addressing one of the major concerns of West Africa, and Burkina Faso in particular, namely "How to improve pasture productivity for sustainable livestock production in a context of security, climate and land crises".

It would then be advisable to continue the study of plant invasion in the sense of :

- map the distribution of *S. obtusifolia* in the country's pastoral zones,
- test the germination capacity of *Senna obtusifolia* seeds according to storage time to determine the seed's resistance capacity in terms of shelf life,
- Extend set-aside in the Sudano-Sahelian and Sudanian zones
- identify ecotypes of *Senna obtusifolia* to compare its productivity and adaptation to environmental conditions, particularly water and heat stress.
- Propose remote sensing monitoring and modeling of the evolution of invasive species in pastoral areas.
- Evaluate the nutrient composition of *S. obtusifolia* dry stems for use as a feed supplement.

Recommendations

For the management of invasive species and degraded pastures, the following recommendations have been made:

- For local populations, a rotational farming system including set-asides within pastoral zones to enable restoration of services in degraded areas. They should also carry out soil restoration activities to accelerate their recovery;
- The authorities in charge of the environment need to pay particular attention to the problem of invasive species, in order to set up systems for their prevention and sustainable management. It is also important to put in place a system for securing pastoral areas through rigorous monitoring of these zones to improve the production of the livestock system.
- There is a need to promote the use of dry *S. obtusifolia* stems in livestock feed.
- Promote the cultivation of natural fodder throughout the country to make it available.

References

Abakar, I., Tankoano, A., & Aly, S. (2019). Traditional Technologies and Probiotic Properties of Bacillus Strains Isolated from Kawal -A Chad Traditional Fermented Food Condiment. *Journal of Food Technology Research*, 6(2), 57-71. https://doi.org/10.18488/journal.58.2019.62.57.71

Abbott, T. P., Vaughn, S. F., Dowd, P. F., Mojtahedi, H., & Wilson, R. F... (1998). Potential uses of sicklepod (cassia obtusifolia). *Crops and industrial products*, 8 (1), 77 -82.

Aboh, A. B. (2008). *Phytosociology, ecology, potentialities and management of natural pastures invaded by Chromolaena odorata and Hyptis suaveolens in the Soudano-Guinean Zone* (Benin) . Doctorat en Gestion des Ressources Naturelles, Aménagement du Territoire et Politique Environnementale, Université d'Abomey-Calavi (UAC), 227 p.

Aboh, A. B., Babatounde, S., Oumorou, M., Houinato, M., & Sinsin, B. (2012). Pastoral value of natural rangelands in Sudano-Guinean zone and peasant strategy of adaptation to the effects of their invasion by Chromolaena odorata in Benin. *International Journal of Biological and Chemical Sciences*, 6(4), 1633-1646.

Abourouh, M., Taleb, M., Makhloufi, M., Boulmane, M., & Aronson, J. (2005). Biodiversity and vegetation dynamics in the Maâmora subéraie (Morocco), effect of enclosure duration. *Forêt méditerranéenne*, 26(4), 275-286.

ACTA (1986). *Cassia obtusifolia* L. In: Adventices Tropicales [Tropical Weeds]. Paris, France: *ACTA-Publications*.

Adams, V. M., & Setterfield, S. A. (2013). Estimating the financial risks of Andropogon gayanus to greenhouse gas abatement projects in northern Australia. *Environmental Research Letters*, 8(2), 025018.

Adjatin A. (2006) Contributions *à l'étude de la diversité des légumes feuilles traditionnels consommés dans le départe-ment de l'Atacora (Benin)*. Université de Lomé (Togo)- Diplôme d'Études 303-Approfondies 2006, 55P.

Agúndez, D., Nouhoheflin, T., Coulibaly, O., Soliño, M., & Alía, R. (2020). Local Preferences for Shea Nut and Butter Production in Northern Benin: Preliminary Results. *Forests*, 11(1), 13. https://doi.org/10.3390/f11010013

Akodéwou A., Oszwald J., Akpavi S., Gazull L., Akpagana K. & Gond V. (2019). Invasive plant issues in southern Togo (West Africa): contribution of landscape systems analysis and remote sensing. *Biotechnology Agrononimy Sociology Environment*, 23 (2): 1-16

Akpo É. L., & Masse D., M. G. (2000). Pastoral value of herbaceous vegetation in Sudanian fallows. *La Jachère En Afrique Tropicale*, 101(January 2016), 493-502.

Akpo L.E., & Grouzis M., (2000). Pastoral value of grasslands in Sudanian regions: the case of Sahelian rangelands in northern Senegal. *Tropicultura*, 18 (1): 1-8

Akpo L.E., Masse D. & Grouzis M. (2002). Fallow duration and pastoral value of herbaceous vegetation in the Sudanian zone of Senegal. *Revue d'élevage et de Médecine Vétérinaire Des Pays Tropicaux*, 55(4), 275. https://doi.org/10.19182/remvt.9815.

Alexiades, M. N., & Sheldon, J. W. (1996). Selected guidelines for ethnobotanical research: A field manual.

Alhassane, A., Soumana, I., Chaibou, I., Karim, S., & Mahamane, A. (2018). Productivity, pastoral value and carrying capacity of rangelands in the Maradi region, Niger Productivity , pastoral value and carrying capacity of rangeland in the Maradi region , Niger. *International Journal of Biolgical and Chemical Sciences*, 12(August), 1705-1716.

Amégnaglo K.B., Dourma M., Akpavi S., Akodewou A., Wala K., Diwediga B., Atakpama W., Agbodan K.M.L., Batawila K., & Akpagana K., (2018). Characterization of grazed plant formations in the Guinean zone of Togo: typology, biomass assessment, diversity, forage value and regeneration. *International Journal of Biology. Chemestry. Sciences, 12* (5): 2065-2084, doi: 10.4314/ijbcs.v12i5.9

Anastasia, A.-a., Awono, K. D., Clautilde, M., & Emmanuel, Y. (2014). Potentials of native herbaceous species for improving country people cultivating systems in the Sudano-Sahelian zone of Cameroon. *Journal of agriculture and environmental sciences*, 3 (3), 73 -89.

APG IV (Angiosperm Phylogeny Group classification for the orders and families of flowering plants). (2016). *Botanical Journal of the Linnean Society*, 181(1): 1-20.

Arbonnier M., 2019. *Arbres, Arbustes et Lianes Des Zones Sèches d'Afrique de l'Ouest.* Paris: MNHN; 776 p.

Archibald, J. L., Anderson, C. B., Dicenta, M., Roulier, C., Slutz, K., & Nielsen, E. A. (2020). The relevance of social imaginaries to understand and manage biological invasions in southern Patagonia. *Biological Invasions*, 22(11), 3307-3323.

Assogbadjo, A. E., Kakaï, R. G., Chadare, F. J., Thomson, L., Kyndt, T., Sinsin, B., & Van Damme, P. (2008). Folk classification, perception, and preferences of baobab products in West Africa: Consequences for species conservation and improvement. *Economic botany,* 62(1), 74-84.

Attia F., (2007). *Effect of water stress on the ecophysiological behavior and phenological maturity of grapevine (Vitis vinifera L.): Study of five autochthonous grape varieties from Midi-Pyrenées.* Thesis INP, Toulouse (France), 194p.

Augustine, C., Khobe, D., Madugu, A. J., Babakiri, Y., Joel, I., John, T., Igwebuike, J. U., & Ibrahim, A. (2020). Productive performance and cost benefits of feeding wistar albino rats with processed tropical sickle pod (*Senna obtusifolia*) leaf meal-based diets. *Translational Animal Science*, 4(2), 589-593. https://doi.org/10.1093/tas/txaa036

Ayantunde, A. A., Oluwatosin, B. O., Yameogo, V., & van Wijk, M. (2020). Perceived benefits, constraints and determinants of sustainable intensification of mixed crop and livestock systems in the Sahelian zone of Burkina Faso. *International Journal of Agricultural Sustainability*, 18(1), 84-98.

Aziadekey, M. K., Atayi, A., Odah, K., & Magamana, A.-E. (2014). Study of the influence of water stress on two lines of niebe. *European Scientific Journal*, 10(30). https://doi.org/10.19044/esj.2014.v10n30p%p

Baghdadi, A., Halim, R. A., Othman, R., Yusof, M. M., & Atashgahi, A. R. M. (2016). Productivity, relative yield and plant growth of forage corn intercroppedwith soybean under different crop combination ratio. Legume Research - *An International Journal*, 39(4), 558-564. https://doi.org/10.18805/lr.v0iOF.10755

Bailey, R. L., Faulkner-Grant, H. A., Martin, V. Y., Phillips, T. B., & Bonter, D. N. (2020). Nest usurpation by non-native birds and the role of people in nest box management. Conservation Science and Practice, 2(5), e185.Situation and determinants. *Forages*, 221, 15-24.

Balay, C., Cournut, S., Michelin, Y., Capitaine, M., & Boisdon, I. (2015). Ecosystem services rendered by grasslands in a mid-mountain commune of Auvergne: assessment and determinants. *Fourrages*, *221*, 15-24.

Ball, K. R., Power, S. A., Brien, C., & Woodin, S., (2020). High-throughput, image-based phenotyping reveals nutrient-dependent growth facilitation in a grass-legume mixture. *PloS one,* 15(10), e0239673. https://doi.org/10.1371/journal.pone.0239673

Bargali, K., & Bargali, S. (2016). Germination capacity of seeds of leguminous plants under water deficit conditions: Implication for restoration of degraded lands in Kumaun Himalaya. *Tropical Ecology*, 57(3), 445-453.

Barros, V., Melo, A., Santos, M., Nogueira, L., Frosi, G., & Santos, M. G. (2020). Different resource-use strategies of invasive and native woody species from a seasonally dry tropical forest under drought stress and recovery. *Plant Physiology and Biochemistry*, 147, 181-190. https://doi.org/10.1016/j.plaphy.2019.12.018

Baruch, Z. (1994). Responses to drought and flooding in tropical forage grasses. *Plant and Soil*, 164(1), 87. https://doi.org/10.1007/BF00010114

Baskin J. M. & Baskin C. C., (2004). A classification system for seed dormancy. *Seed Science Research*, 14:1-16.

Bastiani, M. O., Lamego, F. P., Agostinetto, D., Langaro, A. C., Silva, D. C. da, Bastiani, M. O., Lamego, F. P., Agostinetto, D., Langaro, A. C., & Silva, D. C. da. (2016). Relative competitiveness of soybean cultivars with barnyardgrass. *Bragantia*, 75(4), 435-445. https://doi.org/10.1590/1678-4499.412

Batawila, K., Akpavi, S., Wala, K., & Kanda, M. (2005). Diversity and management of pick-your-own vegetables in Togo. *Developing African leafy vegetables for improved nutrition*, 55-68.

Bechir A. B. & Mopate L. Y. (2015). Analysis of grazing dynamics around hydraulic structures in pastoral areas of Western Batha, Chad. Afrique Science : *Revue Internationale Des Sciences et Technologie*, 11(1), 212-226.

Becker T & Muller J.V., (2007). Floristic affinities, life-form spectra and habitat preferences of the vegetation of two semi-arid regions in Sahelian West an Southern Africa. *Basic and Applied Dryland Research,* 1:33-50.

Bentham G (1871*). Revision of the genus Cassia*. Trans Linnean Soc London 27:503-591

Berhaut J., (1967). *Flore du Sénégal*. 2nd ed. Clairafrique, Dakar, Senegal, 625p.

Berhaut J., (1979). *Flore illustrée du Sénégal*, Dicotlédones, Dakar, 625 p.

Bewley D. J., Bradford J. K., Hilhorst M. W. H. & Nonogaki H., (2012). Seeds: Physiology of development, germination and dormancy. *Third Edition Springer science*, New york. 392p.

Binggeli P., (2005). *Crop Protection Compendium - Senna obtusifolia* (L.) Irwin & Barneby, 15p.

Black, R., & Bartlett, D. M. (2020). Biosecurity frameworks for cross-border movement of invasive alien species. *Environmental Science & Policy*, 105, 113-119.

Bolinder, M. A., Angers, D. A., Bélanger, G., Michaud, R., & Laverdière, M. R. (2002). Root biomass and shoot to root ratios of perennial forage crops in eastern Canada. *Canadian Journal of Plant Science*, 82(4), 731-737. https://doi.org/10.4141/P01-139

Bonnet P., Arbonnier M., & Grard P., (2005). Ligneux du Sahel. - Cirad - CD-Rom.

Botoni Liehoun, E., Daget, P., & César, J. (2006). Grazing activities, biodiversity and pastoral vegetation in the western zone of Burkina Faso. *Revue d'élevage et de Médecine Vétérinaire Des Pays Tropicaux*, 59, 31-38.

Boudet G., (1991). *Manuel sur les pâturages tropicaux et les cultures fourragères*. Ministère de la coopération et du développement, Coll. Manuels et précis d'élevage 4th ed. Paris France, 266 pages.

Boyette, C. D. (2006). Adjuvants increase the biological control potential of a colletotrichum gloeosporioides isolate for biological control of sicklepod (senna obtusifolia). *Science of Biocontrol and Technology*, 16 (10), 1057-1066.

Bradshaw, C.J.A., & Courchamp, F., (2021). High and rising economic costs of biological invasions worldwide. *Nature* 592, 571-576. https://doi.org/10.1038/s41586-021-03405-6

Brenan JPM (1958) *New and noteworthy cassias from tropical Africa*. Kew Bull 13:231-252

Brenan JPM. (1967). Leguminosae Subfamily Caesalpinioideae. In: Milne-Redhead E, Polhill RM, editors. *Flora of tropical East Africa.* London: White Friars Press, Crown Agents for Oversea Governments and Administrations; p. 230.

Brannen, J. (2005). Méthodes De Mélange: L'entrée des approches qualitatives et quantitatives dans le procédé de recherches. *International Journal of Social Research Methodology*, 8 (3), 173-184.

Bridgewater P, Higgs ES, Hobbs RJ, & Jackson SJ (2011) *Engaging with novel ecosystems. Frontiers in Ecology Environment* 9:423-423

Braun-Blanquet J., (1982). *Plant sociology. The study of plant communities*. First Ed. McGraw-Hill Book Co, New York and London, 439 p.

Broadbent, A., Stevens, C. J., Peltzer, D. A., Ostle, N. J., & Orwin, K. H. (2018). Belowground competition drives invasive plant impact on native species regardless of nitrogen availability. *Oecologia*, 186(2), 577-587.

Buchanan, G. A., Crowley, R. H., Street, J. E., & McGuire, J. A. (1980). Competition of Sicklepod (Cassia obtusifolia) and Redroot Pigweed (Amaranthus retroflexus) with Cotton (Gossypium hirsutum). *Weed Science*, 28(3), 258-262. https://doi.org/10.1017/S0043174500055259

Buchholz J., Querner P., Paredes D., Bauer T., Strauss P., Guernion M., Scimia J., Cluzeau D., Bure F., Kratschmer S., Winter Silvia, Potthoff M. & Zaller J. G. (2017). Soil biota in vineyards are more influenced by plants and soil quality than by tillage intensity or the surrounding landscape. *Scientific reports*, (June), 1-12. https://doi.org/10.1038/s41598-017-17601-w.

Buldgen A., & Dieng A., (1997). Andropogon gayanus var. bisquamulatus. A forage crop for tropical regions. *Les presses agronomiques de Gembloux*, A.S.B.L.

Bunasols (Bureau National des Sols), (1997). *Etude morpho-pédologique de la province du Kénédougou.* Technical report N°108, 172 p.

Bunasols (Bureau national des sols), 2006. *Morpho-pedological study of the Mouhoun and Balés provinces.* Technical report N°135, 82 p.

Burgess, N., Hales, J., Underwood, E., Dinerstein, E., Olson, D., Itoua, I., Schipper, J., Ricketts, T., & Newman, K. (2004). Terrestrial ecoregions of Africa and Madagascar: A conservation assessment. *Island Press*.

Burrows, G. E., & Tyrl, R. J. (2013). Toxic plants of north America. John Wiley & Sons.

Cai, X. & Gu, M. 2016. Bioherbicides in organic horticulture. *Horticulturae* 2:3.

Carey, M. P., Sanderson, B. L., Barnas, K. A., & Olden, J. D. (2012). Native invaders-challenges for science, management, policy, and society. *Frontiers in Ecology and the Environment*, 10(7), 373-381.

Chafai, (2012). *Study of the effect of water stress on a collection of Medicago truncatula lines. Magister en biotechnologies végétales option : Biotechnologies végétales.* Ecole nationale supérieure Agronomique El-harrach Alger, 83p.

Chambers, J. C., Bradley, B. A., Brown, C. S., D'Antonio, C., Germino, M. J., Grace, J. B., Hardegree, S. P., Miller, R. F., & Pyke, D. A. (2014). Resilience to Stress and Disturbance, and Resistance to Bromus tectorum L. Invasion in Cold Desert Shrublands of Western North America. *Ecosystems*, 17(2), 360-375. https://doi.org/10.1007/s10021-013-9725-5

Charahabil, M. M., & Akpo, L. E. (2018). The red guava tree: Psidium cattleyanum var. coriaceum (Mart. ex O. Berg) Kiaersk. An invasive, multipurpose species in Grande-Comore.*ESJ*,14, 436. https://doi.org/10.19044/esj.2018.v14n9p436

Chauvel B., (2019). *When invasive plants also wander into fields.* Encyclopedy. Environment. [online]

Chaves M M., (2002). How plants cope with water stress in the field? Photosynthesis and growth. *Annal of Botany*, 89(7), 907-916.

Chaves M M., Maroco J P., & Pereira J S., (2003). Understanding plant responses to drought: from genes to the whole plant. *Function Plant Biology*. 30. pp : 239-264.

Chaves Neto, J. R., Mazutti, M. A., Zabot, G. L., & Tres, M. V. (2020). Bioherbicidal action of Phoma dimorpha fermented broth on seeds and plants of *Senna obtusifolia*. *Pesquisa Agropecuária Tropical*, 50.

Cissé M., Bationo B. A., Traoré S. & Boussim I. J. (2018). Perception of agroforestry species and their ecosystem services by three ethnic groups in the Boura watershed, Sudanian zone of Burkina Faso. *Bois et Forêts des Tropiques*, no. 338: 29-42. Doi : https://doi.org/10.19182/bft2018.338.a31680.

Côme D., (1993). *"Rôle des facteurs du milieu dans la germination et la survie des semences". In "Les problèmes de semences forestières notamment en Afrique"*. Final proceedings of the symposium of the IUFRO Working Group P.2.04.00. Problèmes des semences, Ouagadougou, Burkina-Faso, November 23-28, 1992, L. M. Somé and M. Kam (Eds). Backhuys Publishers, Leiden, The Netherlands, pp.131-142.

Conedera, M., Bomio-Pacciorini, N., Bomio-Pacciorini, P., Sciacca, S., Grandi, L., Boureima, A., & Vettraino, A. M. (2010). Reconstitution of degraded Sahelian ecosystems. *Bois & forets des tropiques,* 304, 61-71.

Connolly, J., Goma, H. C., & Rahim, K. (2001). The information content of indicators in intercropping research. *Agriculture, Ecosystems & Environment,* 87(2), 191-207. https://doi.org/10.1016/S0167-8809(01)00278-X

Conseil national de l'environnement pour un développement durable, (2010). *Contribution à l'état des connaissances de quelques plantes envahissantes au Niger,* 40p.

Convention on Biological Diversity, (2014). *Fifth national report of Burkina Faso to the conference of the parties to the convention on biological diversity. Burkina Faso.* 114p.

Cordeau, S., Triolet, M., Wayman, S., Steinberg, C., & Guillemin, J.P., (2016). Bioherbicides: Dead in thewater? A review of the existing products for integrated weed management. *Crop Protection.* 87:44-49.

Coton, M. & De C. (1996). *Ethnobotany: Principles and applications.* John Wiley & Sons, New York.

Crawford L, McDonald GM, & Friedman M (1990) Composition of sicklepod (Cassia obtusifolia) toxic weed seeds. *Journal of Agriculture and Food Chemestry* 38:2169-2175

Creel, J. M., Hoveland C. S., & Buchaman G. A... (1968). Germination, growth, and ecology of sicklepod. *Weed Sciences.* 16:396-400.

Cunningham A., B. 1993. African medicinal plants: Setting priorities at the inter-face betwen conservation and primary healthcare. In People and Plants Working Paper. UNESCO.

Daget, P., & Poissonnet, J. (1971). *Principles of a technique for the quantitative analysis of the vegetation of herbaceous formations.* Doc. Cepe-cnrs, 56, 85-100.

Dagnelie, P. (1998). *Theoretical and applied statistics (Volume 2).* De Boeck & Larcier, Paris-Brussels.

Dansi, A., Adjatin, A., Adoukonou-Sagbadja, H., Faladé, V., Yedomonhan, H., Odou, D. &, etDossou, B. (2008). Traditional leafy vegetables and their use in the Benin Republic. *Genetic Resources and Crop Evolution,* 55(8), 1239-1256. https://doi.org/10.1007/s10722-008-9324-z

D'Antonio C. M., T. L. Dudley & M. C. Mack, (1999). Disturbance and biological invasions: direct effects and feedbacks. In: L. Walker (eds.), *Ecosystems of disturbed ground.* pp. 413-452. Elsevier, Amsterdam

Davies, K. W., & Sheley, R. L. (2007). A conceptual framework for preventing the spatial dispersal of invasive plants. *The Science of Weeds,* 55, 178 -184.

DeBoer, M. L., Grev, A. M., Sheaffer, C. C., Wells, M. S., & Martinson, K. L. (2020). Herbage mass, botanical composition, forage nutritive value, and preference of grass-legume pastures under horse grazing. *Crop, Forage & Turfgrass Management*, 6(1), 20032.

Dekker, J. H., Meggitt, W. F., & Putnam, A. R. (1983). Experimental methodologies to evaluate allelopathic plant interactions. *Journal of Chemical Ecology*, 9(8), 945-981. https://doi.org/10.1007/BF00982204

Department of Agriculture and Fisheries, (2016). *Restricted invasive plant: Sicklepod (Senna obtusifolia), foetid senna (Senna tora) and hairy senna (Senna hirsuta)*. The State of Queensland, www.biosecurity.qld.gov.au

De Wit H C D (1955) *Revision of the genus Cassia as occurring in Malaysia*; Webbia 11 197-222

De Wit, C. T. (1960). *On competition* (No. 66.8). Pudoc.

De Wit, C. T., & Van den Bergh, J. P. (1965). Competition between herbage plants. *Journal of Agricultural Science*, 13, 212-221.

DGEAP, (Direction Générale des Espaces et Aménagements Pastoraux). (2011). *Atelier national de réflexion sur les zones pastorales au Burkina Faso* (p. 35).

Diagne, C., Leroy, B., Vaissière, A.-C., Gozlan, R.E., Roiz, D., Jarić, I., Salles, J.-M., Bradshaw, C.J.A., & Courchamp, F., (2021). High and rising economic costs of biological invasions worldwide. *Nature* 592, 571-576. https://doi.org/10.1038/s41586-021-03405-6

Díaz S., Settele J., Brondízio E., Ngo H., Guèze M., Agard J., Arneth A., Balvanera P., Brauman K., Butchart S., Chan K., Garibaldi L., Ichii K., Liu J., Subrmanian S., Midgley G., Miloslavich P., Molnár Z., Obura D., Pfaff A., Polasky S., Purvis A., Razzaque J., Reyers B., Chowdhury R., Shin Y., Visseren-Hamakers I., Wilis K., & Zayas C., (2019). Summary for policymakers of the global assessment report on biodiversity and ecosystem services of the Intergovernmental Science-Policy Platform on Biodiversity and Ecosystem Services (Project Report). *IPBES*

Diez, J. M., D'Antonio, C. M., Dukes, J. S., Grosholz, E. D., Olden, J. D., Sorte, C. J., Blumenthal, D. M., Bradley, B. A., Early, R., Ibáñez, I., Jones, S. J., Lawler, J. J., & Miller, L. P. (2012). Will extreme climatic events facilitate biological invasions? *Frontiers in Ecology and the Environment,* 10(5), 249-257. https://doi.org/10.1890/110137

Dirar HA (1984). Kawal, meat substitute from fermented Cassia obtusifolia leaves. *Economy of Botany* 38:342-349

Doll, J. D., Piedrahíta Castañeda, W., & Argel M, P. J. (1976). Capacidad germinativa de semillas de 32 especies de malezas. *Revista comalfi.*

Dorning M., & Cipollini D., (2006). Leaf and root extracts of the invasive shrub, Lonicera maackii inhibit seed germination of three herbs with no autotoxic effects. *Plant Ecology* 184:287-296

Doughari, J. H., El-Mahmood, A. M., & Tyoyina, I. (2008). Antimicrobial activity of leaf extracts of *Senna obtusifolia* (L). *African Journal of Pharmacy and Pharmacology*, 2(1), 7-13.

Drabo, B., Grell , H., & Poda , A. (2001). Gestion concertée des ressources agropastorales: cas du Sahel Burkinabé.In E. Tielkes, E. Schlecht, & P.Hiernaux, Elevage et gestion de parcoursau Sahel, implications pour ledéveloppement (p. 15). Beuren-Stuttgart,Germany: Verlag Ulrich E. Grauer.

Dunlop, E. A., Wilson, J. C., & Mackey, A. P. (2006). The potential geographic distribution of the invasive weed *Senna obtusifolia* in Australia. *Weed Research*, *46*(5), 404-413.

Egley GH, & Chandler JM, (1978). Germination and viability of weed seeds after 2.5 years in 50-year buried seed study. *Weed Science*, 26(3):230-239.

Egley, G. H. & J. M. Chandler (1983). Longevity of weed seeds after 5.5 years in the Stoneville 50-year buried-seed study. *Weed Sciences*. 31:264- 270.

Eisenhauer, N., & Scheu, S. (2008). Earthworms as drivers of the competition between grasses and legumes. *Soil Biology and Biochemistry,* 40(10), 2650-2659. https://doi.org/10.1016/j.soilbio.2008.07.010

Eklu-Natey R. D. & Balet A., (2012). *Pharma-copée Africaine Dictionary and multilingual Mo-nographies of the Medical potential of African plants*. West Africa. Vol 2. Édition d'en bas- Traditions et Médecines. T& M Geneva ISBN, 978-2-8290-0433-9.

Ekué, M. R., Sinsin, B., Eyog-Matig, O., & Finkeldey, R. (2010). Uses, traditional management, perception of variation and preferences in ackee (Blighia sapida KD Koenig) fruit traits in Benin: Implications for domestication and conservation. *Journal of Ethnobiology and Ethnomedicine*, 6(1), 1-14.

El-Barougy, R., MacIvor, J. S., Arnillas, C. A., Nada, R. M., Khedr, A.-H. A., & Cadotte, M. W. (2020). Richness, phylogenetic diversity, and abundance all have positive effects on invader performance in an arid ecosystem. *Ecosphere*, 11(2), e03045. https://doi.org/10.1002/ecs2.3045

Essl, F., Bacher, S., Genovesi, P., Hulme, P.E., Jeschke, J.M., Katsanevakis, S., Kowarik, I., Kühn, I., Pyšek, P., Rabitsch, W., Schindler, S., van Kleunen, M., Vilà, M., Wilson, J.R.U., & Richardson, D.M., (2018). Which Taxa Are Alien? Criteria, Applications, and Uncertainties. *BioScience* 68, 496-509. https://doi.org/10.1093/biosci/biy057

Ewart A, (1908). *On the longevity of seeds. Proceedings of the Royal Society Victoria*, 21:1-210.

FAO (Food and Agriculture Organization), (1992). *Forest seed handling guide*. Ed. Rome (Italy) (8): 203-235.

FAO, 2019. *The future of livestock farming in Burkina Faso. Challenges and opportunities in the face of uncertainties*. Rome. 56 p. License: CC BY-NC-SA 3.0 IGO.

Ferner, J., Linstädter, A., Südekum, K.-h., & Schmidtlein, S. (2015). Spectral indicators of forage quality in the tropical savannas of West Africa. *International Journal of Applied Earth Observation and Geoinformation*, 41, 99 -106.

Figuiere F., Marnotte P., Le Bourgeois T., & Carrara A., (1998). Determination key for eight species of the genus cassia L. (Caesalpiniaceae), West African weeds. *Agriculture et développement*, n°19-September 1998.

Fischer, R. A., & Maurer, R. (1978). Drought resistance in spring wheat cultivars. I. Grain yield responses. *Australian Journal of Agricultural Research*, 29(5), 897-912. https://doi.org/10.1071/ar9780897

Fontes J., & Guinko S., (1995). *Vegetation and land use map of Burkina Faso*. Explanatory note. Toulouse, France, French Ministry of Cooperation, Campus project, 68 p.

Fournier A., (1983). Contribution to the study of herbaceous vegetation of savannas for Ouango-Fitini (Ivory Coast). *Candollea,* 38:237-265.

Fournier, A. (1991). *Phenology, growth and plant production in some West African savannas: variation along a climatic gradient.*

Fournier, A. (1994). Cycle Saisonnier Et Production Nette. *Ecologie,* 25(3), 173-188.

Fowler, N. (1982). Competition and Coexistence in a North Carolina Grassland: III. Mixtures of Component Species. *Journal of Ecology*, 70(1), 77-92. JSTOR. https://doi.org/10.2307/2259865

Fried G., (2019). *Apports des approches fonctionnelles pour l'évaluation des risques associés aux changements de végétation induits par les activités humaines.* Habilitation to direct research, University of Montpellier.

Funk, J. L., Standish, R. J., Stock, W. D., & Valladares, F. (2016). Plant functional traits of dominant native and invasive species in mediterranean-climate ecosystems. *Ecology,* 97(1), 75-83.

Gaméné C S., (1987). *Contribution à la maîtrise des méthodes simples de prétraitement et de conservation des semences de quelques espèces ligneuses récoltées au Burkina Faso.* Diploma thesis. IDR, University of Ouagadougou, 94 p.

GAMPINE D., (1992). *Etude de la germination et des plantules de quelques essences spontanées de combretaceae et caesalpiniaceae au Burkina Faso,* dissertation. Institut du Développement Rural (IDR)/Université de Ouagadougou, 124p.

Gannouka, N. (2021). *Study of the potential of service plants to improve sorghum (Sorghum bicolor L.) yields.* Joseph Ki-Zerbo University.

García-Llorente, M., Martín-López, B., González, J. A., Alcorlo, P., & Montes, C. (2008). Social perceptions of the impacts and benefits of invasive alien species: Implications for management. *Biological Conservation*, 141(12), 2969-2983.

Gard B., (2012). *Ecological and evolutionary processes influencing the colonization of mugwort ragweed (Ambrosia artemisiifolia L.)* in France. Dijon

Gaoue, O. G., Sack, L., & Ticktin, T. (2011). Human impacts on leaf economics in heterogeneous landscapes: The effect of harvesting non-timber forest products from African mahogany across habitats and climates. *Journal of Applied Ecology*, 48(4), 844-852.

Gebrekiros M. G., & Tessema Z. K. (2018). Effect of *Senna obtusifolia* (L.) invasion on herbaceous vegetation and soil properties of rangelands in the western Tigray, northern Ethiopia. *Ecological Processes*, 7(1). https://doi.org/10.1186/s13717-018-0121-0.

Gebreyesus, G. M. (2017). Livestock herders perception on the causes and effects of *Senna obtusifolia* L. invasion in rangelands of Northern Ethiopia. *African Journal of Agricultural Research,* 12(42), 3081-3087.

Guinko S., (1984). *Végétation de la Haute Volta Thèse de doctorat es sciences naturelles.* Université de Bordeau III. 318 pages.

Glèlè KaKaï R., SALAKO V. K. & Lykke A. M. (2016). Sampling techniques in vegetation survey. *Annales des Sciences Agronomiques* 20 - spécial Projet Undesert-UE: 1-13 (2016) ISSN 1659-5009. Google Earth Pro (GEP).

Godefroid S. & Koedam N., (2003). Distribution pattern of the flora in a peri-urban forest: an effect of the city-forest ecotone. *Landscape and urbain planning*, 65:169-185.

Gomgnimbou, A. P., Bandaogo, A. A., Kalifa, C., Sanon, A., Ouattara, S., & Nacro, H. B. (2019). Short-term effects of poultry droppings application on maize (Zea mays L.) yield and chemical characteristics of a ferralitic soil in the South Sudanian zone of Burkina Faso. *International Journal of Biological and Chemical Sciences*, 13(4), 2041-2052.

Goudard A., (2007). *Ecosystem functioning and biological invasions: the importance of biodiversity and interspecific interactions.* Université Pierre et Marie Curie, Paris VI, France, 216 p.

Goudégnon, E. O. A., Vodouhe, F. G., Gouwakinnou, G. N., Salako, V. K., & Oumorou, M. (2018). Difference between generations and ethnic groups of traditional knowledge and cultural importance of Lannea microcarpa Engl. & K. Krause in Sudanian savanna in Benin | *Bois & Forets des Tropiques.* https://doi.org/10.19182/bft2017.334.a31491

Groot, J. R., Traoré, M., & Koné, D. (1998). Description of the root system of three forage species in the Sudano-Sahelian zone: *Andropogon gayanus*, *Vigna unguiculata* and *Stylosanthes hamata. BASE.*

Grouzis, M. (1984). *Sahelian pastures in northern Burkina-Faso: carrying capacity, frequency of production and forage quality dynamics*, 35p.

Grubben, G. J. H. & Denton, O. A. (2004). Plant resources of tropical Africa 2. Vegetables. [Plant Resources of Tropical Africa 2. Vegetables. 2004]. *PROTA Foundation*, Wageningen, The Netherlands / Backhuys Publishers, Leiden, The Netherlands / CTA, Wageningen, The Netherlands, 737p

Grüner, E., Wachendorf, M., & Astor, T. (2020). The potential of UAV-borne spectral and textural information for predicting aboveground biomass and N fixation in legume-grass mixtures. *PloS one*, *15*(6), e0234703. https://doi.org/10.1371/journal.pone.0234703

Hamilton, A., Shengji, P., Kessy, J. K. A. A., Khan, A. A., Lagos-Witte, S., & Shinwari, Z. K. (2003). *The purposes and teaching of applied ethnobotany* (Vol. 11). United Nations Educational, Scientific and Cultural Organization (UNESCO).

Hartvigsen, G. (2000). Competition between co-dominant plants of the Serengeti plains depends on competitor identity, water, and urine. *Plant Ecology*, 148(1), 31-41. https://doi.org/10.1023/A:1009835424490

Hauser, E. W., Buchanan, G. A., & Ethredge, W. J. (1975). Competition of Florida Beggarweed and Sicklepod with Peanuts I. Effects of Periods of Weed-free Maintenance or Weed Competition. *Weed Science*, 23(5), 368-372. https://doi.org/10.1017/S0043174500062688

He, Y., Liu, X., Wang, M., & Sun, S., (2022). Grazing alters seedling emergence number, dynamics, and diversity of herbaceous plants in a semiarid sandy grassland. *Ecological Research*. https://doi.org/10.1111/1440-1703.12355

Hellmann, J. J., Byers, E. De J., Bierwagen, B. G., & Ducs, S. De J. (2008). Five potential consequences of climate change for invasive species. *Conservation Biology*, 22 (3), 534 -543.

Helsen, K., Smith, S. W., Brunet, J., Cousins, S. A., De Frenne, P., Kimberley, A., ... & Graae, B. J. (2018). Impact of an invasive alien plant on litter decomposition along a latitudinal gradient. *Ecosphere*, 9(1),e02097.https://pics.davesgarden.com/pics/2007/08/21/Farmerdill/7f0406.jpg,

Hiernaux, P., Dardel, C., Kergoat, L., & Mougin, E. (2016). Desertification, adaptation and resilience in the Sahel: Lessons from long term monitoring of agro-ecosystems. *In The end of desertification?* (pp. 147-178). Springer.

Higazi, A., & Abubakar Ali, S. (2018). *Pastoralism and Security in West Africa and the Sahel: Towards peaceful coexistence.*

Hilty J (2018) Illinois wildflowers. *Last updated October* 24, (2018). https://www. illinoiswildflowers.info/prairie/plantx/sicklepodex.htm. Accessed: January22, 2021

Hobbs, R. J., & Huenneke, L. F. (1992). Disturbance, Diversity, and Invasion: Implications for Conservation. *Conservation Biology*, 6(3), 324-337.

Holm LG, Pancho JV, Herberger JP, & Plunknett DL (1979) A Geographical Atlas of World Weeds. Malabar, FL: *Krieger Publishing*. 391 p

Holm L, Doll J, Holm E, Pancho J, & Herberger J (1997) World Weeds; Natural Histories and Distribution. New York: *John Wiley & Sons* 1,129 p

Holmes, T. P., Aukema, J. E., Von Holle, B., Liebhold, A., & Sills, E. (2009). Economic impacts of invasive species in forest past, present, and future. In: The Year In Ecology and Conservation Biology, 2009. *Annal. NY Academy. Sciences.* 1162: 18-38., 1162, 18-38. https://doi.org/10.1111/j.1749-6632.2009.04446.x

Holt, J. S. (1995). Plant Responses to Light: A Potential Tool for Weed Management. *Weed Science*, 43(3), 474-482. https://doi.org/10.1017/S0043174500081509

Hopkins G W., (2003). *Plant physiology*, translated from English by rambour S. Edit. De Boeck, pp. 38 ,58 ,458.

Hopkins, (2013). *Plant physiology*, 1st edition; 513p

Hoveland C S., & Buchanan G A., (1973). Weed seed germination under simulated drought. *Weed Science*, 21(4) :322-324.

Idrissou Y., Mama Sambo Seidou Y., Assani Seidou A., Sanni Worogo H.S., Assogba B.G.C., Alkoiret Traoré I., & Houinato M., (2020). Influence of grazing and climatic gradient on the flora diversity and productivity of rangelands in Benin. *Rev. Elevage. Medecine. Veterinary. Pays Trop*, 73 (3): 000-000, doi: 10.19182/remvt.31894

Inkoto, C. L., Nicole, L., Ashande, C. M., Masens, Y. D.-M., & Mpiana, P. T. (2019). Ethnobotanical and floristic study of some medicinal plants marketed in Kinshasa, Democratic Republic of Congo. *Revue Marocaine Des Sciences Agronomiques et Vétérinaires*, 7(1).

INRA France. (2017). *Evaluation of ecosystem services provided by agricultural ecosystems. Agricultural ecosystems component of the French Evaluation of Ecosystems and Ecosystem Services.* Study report, 966 p.

INRA (2006). *Sécheresse et agriculture: réduire la vulnérabilité de l'agriculture à un risque accru de manque d'eau.* Synthèse du rapport d'expertise scientifique collective, 76p.

INSD (Institut National de la Statistique et de la Démographie), (2020). *ÉtudeMonographique sur la Démographie, la Paix et la sécurité Au Sahel Projections démographiques 2007-2020* 407p.

IPBES (2019). *Global assessment report on biodiversity and ecosystem services of the Intergovernmental Science-Policy Platform on Biodiversity and Ecosystem Services.* Zenodo. https://doi.org/10.5281/Zenodo.3831673

Irwin HS, & Barneby RC (1982). The American Cassiinae. New York: New York *Botanical Garden* 35:1-918

Issaharou-Matchi, I., Barboni, D., Meunier, J.-D., Saadou, M., Dussouillez, P., Contoux, C., & Zirihi-Guede, N. (2016). Intraspecific biogenic silica variations in the grass species Pennisetum pedicellatum along an evapotranspiration gradient in South Niger. *Flora - Morphology, Distribution, Functional Ecology of Plants*, 220, 84-93. https://doi.org/10.1016/j.flora.2016.02.008

ISTA (International Seed Testing Association), (2009). *International rules for seeds testing. Seed Sciences and Technology.* Volume 27.1-10p.

Jabran, K.; Mahajan, G.; Sardana, V.; & Chauhan, B. S. Allelopathy for weed control in agricultural systems. *Crop Protection*, v. 72, n. 1, p. 57- 65, (2015).

Johnson D. E., (1997). *Weeds in rice cultivation in West Africa.* CTA/WARDA, 312p.

Jeangros, B., & Berthola, C. (2002). Long-term evolution of an intensively managed meadow after cessation of fertilization and reduction of cutting frequency. In Multi-function grasslands: quality forages, animal products and landscapes. Proceedings of the 19th General Meeting of the European Grassland Federation, La Rochelle, France, 27-30 May 2002 (pp. 794-795). *Organizing Committee of the European Grassland Federation.*

Kagambèga, W. F., Nana, R., Bayen, P., Thiombiano, A., & Issaka Boussim, J. I. (2019). Water deficit tolerance of five priority species for reforestation in Burkina Faso. *BASE.* https://doi.org/10.25518/1780-4507.18199

Kadeba, A., Nacoulma, M. I. De B., Ouédraogo, A., Thiombiano, A., & Boussim, I. J. (2015). Landcover change and Sahel plant diversity: a case study from northern Burkina Faso. *Annals of forest research*, 58 (1), 109 -123.

Kelso, M. A., Wigginton, R. D., & Grosholz, E. D. (2020). Nutrients mitigate the impacts of extreme drought on plant invasions. *Ecology,* 101(4), e02980. https://doi.org/10.1002/ecy.2980

Keuskamp, J., Dingemans, B. J. J., Lehtinen, T., Sarneel, M. De J., & Soulever, M. De M. (2013). Index De Sachet à thé: A novel approach to gather uniform decomposition data across ecosystems. *Methods in Ecology and Evolution*, 4 (11), 1070 -1075.

Khatiwada, B., Acharya, S. N., Larney, F. J., Lupwayi, N. Z., Smith, E. G., Islam, M. A., & Thomas, J. E. (2020). Benefits of mixed grass-legume pastures and pasture rejuvenation using bloat-free legumes in western Canada: A review. *Canadian Journal of Plant Science*, 100(5), 463-476.

Kiema A. (2002). *Ressources pastorales et leurs modes d'exploitation dans deux terroirs sahéliens du Burkina Faso*. DEA dissertation, Université Polytechniue de Bobo-Diolasso, 65p.

Kiema S. (2007). *Extensive livestock farming and biodiversity conservation in protected areas of western Burkina Faso. A look at their history, current management challenges, vegetation status and dynamics*. University of Orléans thesis, 657p.

Kiema, A., Nianogo, A., J., Somda, J., & Ouédraogo, T. (2008). Valorization of *Cassia obtusifolia* L. in feed for fattening sheep in the Sahelian region of Burkina Faso. *Tropicultura,* 2 (2), 98-103.

Kiema A., Nianogo A. J., kabore-Zoungrana C. Y. & Jalloh B. (2012). Effects of half-moons associated with scarification on forage production in the Sahelian region of Burkina Faso. *International Journal of Biological and Chemical Sciences*, 6(December), 4018-4030. https://doi.org/http://dx.doi.org/10.4314/ijbcs.v6i6.13.

Kiema, A., Tontibomma, G., & Zampaligré, N. (2014). Transhumance and natural resource management in the Sahel: Constraints and prospects in the face of changes in pastoral production systems. *VertigO: la revue électronique en sciences de l'environnement*, 14(3).

Koura, K., Ganglo, J. C., Assogbadjo, A. E., & Agbangla, C. (2011). Ethnic differences in use values and use patterns of Parkia biglobosa in Northern Benin. *Journal of ethnobiology and ethnomedicine*, 7(1), 1-12.

Cousens, R. & O'Neill, M. (1993) Density dependence of replacement series experiments. *Oikos*, 66 (2), 347-352.

Krebs, C., Gerber, E., Matthies, D., & Schaffner, U. (2011). Herbivore resistance of invasive Fallopia species and their hybrids. *Oecologia,* 167(4), 1041-1052.

Kueffer, C. (2017). *Plant invasions in the Anthropocene*. Science, 358(6364), 724-725. https://doi.org/10.1111/ddi.12680

Kumar, M., Singh, H., & Singh, V. (2017). Competitive Interactions between Rice and Caesulia axillaris or Echinochloa crus-galli: A Replacement Series Study. *Annals of Agricultural & Crop Sciences*, 2(1). https://www.google.com

Kumar, M., Padalia, H., Nandy, S., Singh, H., Khaiter, P., & Kalra, N. (2020). Does spatial heterogeneity of landscape explain the process of plant invasion? A case study of Hyptis suaveolens from Indian Western Himalaya. *Environmental Monitoring and Assessment*, 191(3), 794. https://doi.org/10.1007/s10661-019-7682-y

Lacroix P., Magnanon S., Geslin J., Hardegen M., Le Bail J. & Zambettakis C. 2007. Les plantes invasives des régions Basse-Normandie, Bretagne et Pays de la Loire. Définitions et clé pour l'élaboration de listes de " plantes invasives ", " potentiellement invasives " ou " à surveiller ", Conservatoire Botanique National de Brest, Document technique Connaissance de la flore, 12 pp. + annexes.

Leal, R. P., Silveira, M. J., Petsch, D. K., Mormul, R. P., & Thomaz, S. M. (2022). The success of an invasive Poaceae explained by drought resilience but not by higher competitive ability. *Environmental and Experimental Botany*, 194, 104717.

Leßmeister, A., Bernhardt-Römermann, M., Schumann, K., Thiombiano, A., Wittig, R., & Hahn, K. (2019). Vegetation changes over the past two decades in a West African savanna ecosystem. *Applied Vegetation Science*, 22(2), 230-242. https://doi.org/10.1111/avsc.12428

Le Bourgeois T., (2008). *Invasive plants, a major constraint of tropical crops: knowing weeds for sustainable management of cultivated or natural environments.* In: Canne à sucre. CIRAD, Montpellier, 2

Legese, B., & Balew, A. (2021). Land-use and land-cover change in the lowlands of Bale Zone, Ethiopia: Its driving factors and impacts of rangeland dynamics in livestock mobility. *Environmental Monitoring and Assessment*, 193(7), 1-17.

Levine, J. M., Adler, P. B., & Yelenik, S. G. (2004). A meta-analysis of biotic resistance to exotic plant invasions. *Ecology Letters*, 7, 975-989.

Linders, T. E., Bekele, K., Schaffner, U., Allan, E., Alamirew, T., Choge, S. K., & Eschen, R. (2020). The impact of invasive species on social-ecological systems: relating supply and use of selected provisioning ecosystem services. *Ecosystem services*, *41*, 101055. https://doi.org/10.1016/j.ecoser.2019.101055

Tjaden, W. L. (1981). Amaryllis belladonna L. Species plantarum 293, 1753. *Taxon*, *30*(1), 294-298.

Linnaeus C, (1753). *Species Plantarum edition* 1. Stockholm, Sweden. Volume 67 P. 189- 212

Law n°034-2002/an on the orientation law relative to pastoralism in Burkina Faso.

Lowe S., Browne M., Boudjelas S., & De Poorter M., (2007). 100 of the world's most damaging invasive alien species: a selection from the global invasive species database. Premiere. International Union for Conservation of Nature (IUCN), *Gland, Switzerland*, 12 p.

M. F. Obulbiga, C. A. T. Gava, V. Bougouma, H. O. Sanon, D. Kocty, D. Kiemde, A. Coulibaly & M. Diallo, (2018). Effect of cuttings type of fodder palm (Opuntia ficus-indica (L.) Mill) on resprouting capacity and plant development under rainfed conditions in Burkina Faso. *International. Journal of Biology and Chemestry Sciences.* 12(3): 1199-1207.

M. Koutou, M. Sangaré1, M. Havard, E. Vall, L. Sanogo, T. Thombiano, & D. S. **Vodouhe** (2016). Adaptation of livestock practices by producers in western Burkina Faso in the face of land and health constraints. *Agronomie Africaine* 28 (2): 13 - 24 (2016).

Mackey PA, Miller EN, & Palmer WA (1997) *Sicklepod (Senna obtusifolia) in Queensland-pest status review series land protection.* Department of Natural

Resources, Queensland Retrieved on February 29/2015 from file/0011/63875/IPA-Sicklepod-PSA.

Mackey A.P., E.N. Miller, & W. A. P. (2014). Sicklepod (*Senna obtusifolia*) in Queensland. *Pest Status Review Series* - Land Protection, (September), 42p.

Maestre, F. T., Eldridge, D. J., Soliveres, S., Kéfi, S., Delgado-Baquerizo, M., Bowker, M. A., Garcia-Palacios, P., Gaitan, J., Gallardo, A., Lazaro, R., & Berdugo, M. (2016). Structure and functioning of dryland ecosystems in a changing world. Tour. *Ecology. Evolution. System.* , 47, 215 -237.

Magurran A.E., (2004). *Measuring biological diversity.* A primer on methods and computing. Jhon Willey & Sons, 1988, Inc.pp 67-103.

Marazzi B, Endress PK, Paganucci De Queiroz L, & Conti E (2006) Phylogenetic relationships within Senna (Leguminosae, Cassiinae) based on three chloroplast DNA regions: patterns in evolution of floral symmetry and extrafloral nectaries. *American Journal of Botany* 93:288-303

Maregesi, S. M., Ngassapa, O. D., Pieters, L., & Vlietinck, A. J. (2007). Ethnopharmacological survey of the Bunda district, Tanzania: Plants used to treat infectious diseases. *Journal of ethnopharmacology*, 113(3), 457-470.

Martin, M. P. L. D., & Field, R. J. (1984). The nature of competition between perennial ryegrass and white clover. *Grass and Forage Science*, 39(3), 247-253. https://doi.org/10.1111/j.1365-2494.1984.tb01689.x

Martin, G. D., & Coetzee, J. A. (2014). Competition between two aquatic macrophytes, Lagarosiphon major (Ridley) Moss (Hydrocharitaceae) and Myriophyllum spicatum Linnaeus (Haloragaceae) as influenced by substrate sediment and nutrients. *Aquatic Botany*, 114, 1-11. https://doi.org/10.1016/j.aquabot.2013.11.001

Masengo, C., Bongo, G., Robijaona, B., Ilumbe, G., Koto, J.-P. N., & Mpiana, P. (2021). Quantitative ethnobotanical study and sociocultural value of Lippia multiflora Moldenke (Verbenaceae) in Kinshasa, Democratic Republic of Congo. *Revue Marocaine des Sciences Agronomiques et Vétérinaires,* 9(1).

Masson-Delmotte, V., Zhai, P., Pörtner, H. O., Roberts, D., Skea, J., & Shukla, P. R. (2022). *Global Warming of 1.5° C: IPCC Special Report on Impacts of Global Warming of 1.5° C above Pre-industrial Levels in Context of Strengthening Response to Climate Change, Sustainable Development, and Efforts to Eradicate Poverty.* Cambridge University Press.

Maxwell S.L., Fuller R.A., Brooks T.M., &Watson J.E., (2016). Biodiversity: The ravages of guns, nets and bulldozers. *Nature,* 536 (7615): 143, doi: 10.1038/536143

Mbaye N., Diop A T., Gueye M., Diallo A T., Sall C E., & Samb P I., (2002). Study of the germinative behavior and tegumental inhibition lifting tests of Zornia glochidiata Reichb. ex DC. seeds, forage legume. *Revue Élevage. Médecine. Véterinaire Pays Trop*, 55, 1, 47-52.

McGilchrist, C. A., & Trenbath, B. R. (1971). A Revised Analysis of Plant Competition Experiments. *Biometrics*, 27(3), 659-671. JSTOR. https://doi.org/10.2307/2528603

Meekins, J. F., & McCarthy, B. C. (1999). Competitive Ability of Alliaria petiolata (Garlic Mustard, Brassicaceae), an Invasive, Nonindigenous Forest Herb. *International Journal of Plant Sciences,* 160(4), 743-752. https://doi.org/10.1086/314156

MED, (2005). Monographie de la province de la Comoé. Ministry of Economy and Development, Burkina Faso. Report, 126 p + appendices.

Mefti M., Abdelguerfi A., & Chebouti A., (2001). Study of drought tolerance in some populations of Medicago truncatula (L.) Gaertn. in: D elgado I. (ed.), Lloveras J. (ed.). Quality inlucerne and medics for animal production. *Zaragoza*: CIHEA.

Métro A. et Sauvage Ch. Flore des végétaux ligneux de la Mamora. La nature du Maroc I,

1955, 499 pages.

Meuller-Steover, D.; Nybroe, O.; Baraibar, B.; Loddo, D.; Eizenberg, H.; French, K.; Sønderskov, M.; Neve, P.; Peltzer, D. A.; Maczey, N.; & Christensen, S., (2016). Contribution of the seed microbiome to weed management. *Weed Research*, v. 56, n. 5, p. 335-339.

Meyer, S. T., Koch, C., &Weisser, W. W. (2015). Towards a standardized rapid ecosystem function assessment (REFA). *Trends in ecology and evolution*, 30, 390-397.

Millennium Ecosystem Assessment. (2005). *Ecosystems and Human Well-being: Synthesis*. Island Press, Washington, DC, 137p.

Millogo, R. J. (2001). *L'Homme, le climat et les ressources alimen-taires végétales en périodes de crise de subsistance au cours du 20e siècle au Burkina Faso*. University of Ouagadougou.

Ministry of Animal Resources (2010). *Politique nationale de développement durable de l'élevage au Burkina Faso*. Report, 45p.

MEF (2008). Recensement général de la population et de l'habitation de (2006). INSD, July 2008.

Mitja D., & Pascale de Robert, (2004). Innovative farmers: a new grass in the pastures of Santa Maria (Brazilian Amazon). *Natures Sciences Sociétés* 12, 285-298 (2004) NSS-Dialogues, EDP Sciences 2004 ; DOI: 10.1051/nss:2004040

Moleele, M. De N., Ringrose, S., Matheson, W., & Vanderpost, C. (2002). More wooded mills... bush encroachment status in Botswana's grazing regions. *Journal of environmental management*, 64, 3 -11.

Ministry of Animal Resources (MRA). (2016). Ressources Génétiques Animales de État et Au Burkina Faso d Aquacoles. Burkina Faso Report.

Monks, D. W., & Oliver, L. R. (1988). Interactions Between Soybean (Glycine max) Cultivars and Selected Weeds. *Weed Science*, 36(6), 770-774. https://doi.org/10.1017/S0043174500075809

Moreshet, S., Bridges, D. C., NeSmith, D. S., & Huang, B. (1996). Effects of Water Deficit Stress on Competitive Interaction of Peanut and Sicklepod. *Agronomy Journal*, 88(4), 636-644. https://doi.org/10.2134/agronj1996.00021962008800040023x*

Moussa B. M., Soumana I., Saley K., Ambouta K. J-M. & Mahamane A. (2018). Diversity, productivity and quality of pastures in dune ecosystems of Gouré Department (Niger). *Geo-Eco-Trop*, 42, 2: 285-296.

Muller C., (1986). *Forest seed saving and dormancy release.* Technical note. 200-204p.

Muller, S. (2000). Les espèces végétales invasives en France: Bilan des connaissances et propositions d'actions. *Revue d'Ecologie* (La Terre et La Vie), (SUPPL. 7), 53-69.

Nacambo, H., Nanema, K. R., Kiebre, M., Taore, R. E., Sawadogo, N., Ouedraogo, M. H., Tiama, D., Bougma, L. A., & Sombie, P. A. E. D. (2021). Local nomenclature and uses of *Senna obtusifolia* (L.) in Burkina Faso. *Journal of Applied Biosciences*, 160, 16438-16453. https://doi.org/10.35759/JABs.160.1

Nana R., Tamini Z., Sawadogo M. & Some P.P., (2010). Comparative morphological study of five varieties of okra [Abelmoschus esculentus (l.) moench] subjected to water stress.*Journal of .Sciences.*Vol.10, N°3 p 28-38

Nacoulma, M., B., I., Ouedraogo, I., Ouedraogo, O., Dimobe, K., & Thiombiano, A., (2019). Phytodiversity of Burkina Faso in Global Biodiversity.

Ngom D., Bakhoum A., S. D. & L. E. A. (2012). Pastoral quality of grassland resources in the Nord Ferlo Biosphere Reserve (Northern Senegal). International *Journal of Biolgical and Chemical Sciences*, 6(February), 186-201.

Nicolopoulou-Stamati P., Maipas S., Kotampasi C., Stamatis P., & Hens L., (2016). Chemical Pesticides and Human Health: The Urgent Need for a New Concept in Agriculture. *Frontiers in Public Health*, 4:148. doi:10.3389/fpubh.2016.00148

Norsworthy J K., & Oliveira M J., (2006). Sicklepod (*Senna obtusifolia* L.) germination and emergence as affected by environmental factors and seeding depth, *Weed Science,* 54:903-909. 2006, 7p.

Nzigidahera B. (2017). *Situation of invasive species in Burundi. Bujumbura,* consultant's report, 76p.55

Obulbiga, M. F., Gava, C. A. T., Bougouma, V., Sanon, H. O., Kocty, D., Kiemde, D., ... & Diallo, M. (2018). Effect of cuttings type of fodder palm (Opuntia ficus-indica (L.) Mill) on resprouting capacity and plant development under rainfed conditions in Burkina Faso. *International Journal of Biological and Chemical Sciences*, *12*(3), 1199-1207.

Odadi, W. O., Fargione, J., & Rubenstein, D. I. (2017). Vegetation, wildlife, and livestock responses to planned grazing management in an African pastoral landscape. *Land Degradation & Development,* 28(7), 2030-2038.

Oliveira, M. T., Matzek, V., Medeiros, C. D., Rivas, R., Falcão, H. M., & Santos, M. G. **(2014).** Stress Tolerance and Ecophysiological Ability of an Invader and a Native Species in a Seasonally Dry Tropical Forest. *Plos one,* 9(8), e105514. https://doi.org/10.1371/journal.pone.0105514

Orgiazzi A., Bardgett R.D., Barrios E., Behan-Pelletier V., Briones M. J. I., Chotte J-L., De Deyn G.B., Eggleton P., Fierer N., Fraser T., H., K., Jeffery S., Johnson N.C., Jones A., Kandeler E., Kaneko N., L., P., Lemanceau P., Miko L., Montanarella L., Moreira F.M.S., R., K.S., Scheu S., Singh B.K., Six, J., van der Putten W.H., & Wall,

D. H. (2016). Global Soil biodiversity atlas. Luxembourg: *Publications Office of the European Union*, L-2995 Luxembourg, Luxembourg. https://doi.org/10.2788/799182.

Ouédraogo A, Thiombiano A, Hahn- Hadjali K, & Guinko S (2006) Diagnosis of the state of degradation of stands of four woody species in Sudanian zones of Burkina Faso. *Sècheresse* 17: 485-491

Ouédraogo O. (2009). *Phytosociology, dynamics and productivity of the vegetation of Arly National Park (southwest Burkina Faso)*. PhD thesis, Univ. Ouaga, 188 p.

Ouédraogo, A., Lykke, A. M., Lankoandé, B., & Korbéogo, G. (2013). Potentials for Promoting Oil Products Identified from Traditional Knowledge of Native Trees in Burkina Faso. *Ethnobotany Research and Applications*, 11, 071-083. https://doi.org/org/vol11/i1547-3465-11-071.pdf

Ouédraogo, I., Nacoulma, M. I. De B., Hahn, K., & Thiombiano, A. (2014). Assessing ecosystem services based on indigenous knowledge in south-eastern Burkina Faso (West Africa*). International Journal of Biodiversity Science, Ecosystem services & Management*, 10 (4): 313-321.

Ouédraogo, O., Bondé, L., Boussim, J. I., &Linstädter, A. (2015). Caught in a human disturbance trap: Response of tropical savanna trees to increasing land use pressure. *Forest Ecology and Management*, 354 : 68-76

Ouedraogo, I., & Ayantunde, A. A. (2019). Biomass production and management practices in mixed crop-livestock systems in the west African Sahel: Opportunities and constraints.

Ouedraogo, K., Zare, A., Ouedraogo, O., & Anja, L. (2021). Resilience strategies of West Africa pastoralists in reponse to scarce forage resources. *Pastoralism*.

Ouedraogo Mathieu & Thiombiano Taladidia, "Déterminants socio-économiques des défrichements agricoles en zone sud-soudanienne du Burkina Faso", Économie rurale [En ligne], 360 | July-August 2017, online August 15, 2019, accessed April 19, 2019. URL: http://journals.openedition.org/ economierurale/5278; DOI: 10.4000/economierurale.5278

Ouédraogo P., Traoré S., Nacoulma B. M. I., Daboue E., & Bationo B. A., 2021. Dissemination and germination of seeds from cattle faeces in the Sahel of Burkina Faso. *Bois et Forêts des Tropiques*, 350: 15-27. Doi : https://doi.org/10.19182/bft2021.350.a36826

Ouoba-Ima, S. (2018). Socio-demographic characteristics and transhumance dynamics of Fulani herders of the Nouhao in Burkina Faso. *VertigO: la revue électronique en sciences de l'environnement, 18*(2).

Parsons WT, & Cuthbertson EG, (1992). Noxious Weeds of Australia. Melbourne, Australia: Inkata Press.

Pasternak D.L., Woltering A., Nikiema D., Senbeto D., Fatondji J. & Ndjeunga J., (2007). Domestication of Senna ob-tusifolia, an important leafy vegetable for the Sahel. ISHS ActaHorticulturae 752: *International Conference on In-digenous Vegetables and Legumes*. Prospectus for Fighting Poverty, Hun-ger andMalnutrition.https://doi.org/10.17660/ActaHortic.2007.752.50

Palmer, W. A., & Pullen, K. R. (2001). The phytophagous arthropods associated with *Senna obtusifolia* (Caesalpiniaceae) in Mexico and Honduras and their prospects for utilization for biological control. *Biological Control*, 20(1), 76-83

Pardo, J., & VanBuren, R. (2021). Evolutionary innovations driving abiotic stress tolerance in C4 grasses and cereals. *The Plant Cell*, 33(11), 3391-3401. https://doi.org/10.1093/plcell/koab205

Paudel, S., Milleville, A., & Battaglia, L. L. (2018). Responses of Native and Invasive Floating Aquatic Plant Communities to Salinity and Desiccation Stress in the Southeastern US Coastal Floodplain Forests. *Estuaries and Coasts*, 41(8), 2331-2339. https://doi.org/10.1007/s12237-018-0419-2

Peng, X., Li, J., Sun, L., & Gao, Y., (2022). Impacts of water deficit and post-drought irrigation on transpiration rate, root activity, and biomass yield of Festuca arundinacea during phytoextraction. *Chemosphere*, 294, 133842. https://doi.org/10.1016/j.chemosphere.2022.133842

Penning de Vries, F WT, & Djitèye M A., (1982). The productivity of Sahelian pastures. A study of soils, vegetation and exploitation of this natural resource. *Agriculture Research Report*. 918, Pudoc, Wageningen.

Pereira, H. M., Ferrier, S., Walters, M., Geller, G. N., Jongman, R. H. G., Scholes, R. J., ... & Wegmann, M. (2013). Essential biodiversity variables. *Science*, 339(6117), 277-278.

Peres, M. T. L. P., da Silva Cândido, A. C., Bonilla, M. B., Faccenda, O., & Hess, S. C. (2010). Phytotoxic potential of Senna occidentalis and *Senna obtusifolia*. Acta Scientiarum. *Biological Sciences*, 32(3), 305-309.

Phillips, O., Gentry, A. H., Reynel, C., Wilkin, P., & Gálvez-Durand B, C. (1994). Quantitative ethnobotany and Amazonian conservation. *Conservation biology*, 8(1), 225-248.

Pierre Binggeli (2005), Crop Protection Compendium - *Senna obtusifolia* (L.) Irwin & Barneby. P 15

Pindard A., (2000). *The water stress-yield relationship of maize in Bresse: what perspective for spatialization? Using a crop simulator (SnCS).* Engineering thesis. Etablissement National d'Enseignement Supérieur Agronomique de Dijon.

(France), 61 p.

Pitelli, R. L. C. M., & Amorim, L. (2003). Effects of different dew periods and temperatures on infection of *Senna obtusifolia* by a Brazilian isolate of Alternaria cassiae. *Biological Control*, 28(2), 237-242.

Plan communal de de développement de la commune urbaine de Dori (PCD) 2015-2019.Juillet 2015. 125p.

Poilecot P., (1995). The Poaceae of Côte d'Ivoire. Boissiera, Vol 50, 734p.

Poilecot, P. (1999). Poaceae of Niger: description, illustration, ecology, uses.

Porensky, L. M. (2011). When edges meet: interacting edge effects in an African savanna. *Journal of Ecology*, 99(4), 923-934.

Prach, K., & Pyšek, P. (1999). How do species dominating in succession differ from others? *Journal of Vegetation Science*, 10(3), 383-392.

Pratt C.F., Constantine K.L., & Murphy S.T., (2017). Economic impacts of invasive alien species on African smallholder livelihoods. *Global Food Security*, 1431-37, doi: ,10.1016/j.gfs.2017.01.011

Prescott, C. E. (2010). Litter decomposition: What controls it and how can we alter it to sequester more carbon in forest soils? *Biogeochemistry*, 101(December 2010), 133-149. https://doi.org/10.1007/s10533-010-9439-0.

Puydarrieux P., & Devaux J. (2013): *Quelle évaluation économique pour les services écosystémiques rendus par les prairies en France métropolitaine?* Notes et Études Socio-économiques n ° 37, pp. 51-86.

PySek, P. (1995). On the terminology used in plant invasion studies. *Plant invasions*: general aspects and special problems, 71-81.

Pyšek P., Davis M.A., Daehler C.C., &Thompson K., (2004). Plant Invasions and Vegetation Succession: Closing the Gap. Bull. *Ecological. Society of America*, 85 (3): 105-109, doi: 10.1890/0012-9623(2004)85[105:PIAVSC]2.0.CO;2

Pyšek, P., Jarošík, V., Hulme, P. E., Kühn, I., Wild, J., Arianoutsou, M., Bacher, S., Chiron, F., Didžiulis, V., & Essl, F. (2010). Disentangling the role of environmental and human pressures on biological invasions across Europe. *Proceedings of the National Academy of Sciences*, 107(27), 12157-12162. https://doi.org/www.pnas.org/cgi/doi/10.1073/pnas.1002314107

Pyšek P., Pergl J., Essl F., Lenzner B., Dawson W., Kreft H., Weigelt P., Winter M., Kartesz J., Nishino M., Antonova L.A., Barcelona J.F., Cabesaz F.J., Cárdenas D., Cárdenas-Toro J., Castaño N., Chacón E., Chatelain C., Dullinger S., Ebel A.L., Figueiredo E., Fuentes N., Genovesi P., Groom Q.J., Henderson L., Inderjit, Kupriyanov A., Masciadri S., Maurel N., Meerman J., Morozova O., Moser D., Nickrent D., Nowak P.M., Pagad S., Patzelt A., Pelser P.B., Seebens H., Shu W., Thomas J., Velayos M., Weber E., Wieringa J.J., Baptiste M.P., &Kleunen M. van, (2017). Naturalized alien flora of the world: species diversity, taxonomic and phylogenetic patterns, geographic distribution and global hotspots of plant invasion. *Preslia*, 89 (3): 203-274, doi: 10.23855/preslia.2017.203

Pyšek, P., Hulme, P. E., Simberloff, D., Bacher, S., Blackburn, T. M., Carlton, J. T., Dawson, W., Essl, F., Foxcroft, L. C., & Genovesi, P. (2020). Scientists' warning on invasive alien species. *Biological Reviews*, 95(6), 1511-1534.

Radosevich, S. R. (1987). Methods to Study Interactions Among Crops and Weeds. *Weed Technology*, 1(3), 190-198. https://doi.org/10.1017/S0890037X00029523

Rakouth R.B., (1989) *Malagasy leguminosae - Potential for fuel wood and reforestation. (Preliminary results) in tree for development in sub sahara Africa* I.F.S. Kenya, pp. 167-182

Randell BR (1988) *Revision of the Cassiinae in Australia*. I. Senna sect. Chamaefistula. J Adelaide Bot Gard 11:19-49

Randell BR (1995) *Taxonomy and evolution of Senna obtusifolia and S. tora*. J Adelaide Bot Gard 16:55-58

Raunkiaer C., (1934). *The life forms of plants and statistical plants geography*. Clarendron Press, Oxford, 632 p.

RCore Team, (2021). *A language and environment for statistical computing*. Vienna, Austria, R Foundation for Statistical Computing.

Reid, M. De A., Morin, L., Downey, P. O., Français, K. & Vertu, J. G. (2009). Does invasive corporate management facilitate restoration of normal ecosystems? *Biological Conservation*, 142, 2342 - 2349.

Retzinger EJ Jr (1984) Growth and development of sicklepod (Cassia obtusifolia) selections. *Weed Science* 32:608-611

Pitelli R.L.C.M.; & Amorim L., (2003).   Effects of different dew periods and temperatures on infection of *Senna obtusifolia* by a Brazilian isolate of Alternaria cassia. *Biological Control* 28 (2003) 237-242.

Richardson D. M., Pysek P., Rejmanek M., Barbour M. G., F. D. P. & C. J. W. (2000). Naturalization and invasion of alien plants: Concepts and definitions. *Diversity and Distributions*, 6(2), 93-107. https://doi.org/10.1046/j.1472-4642.2000.00083-x.

Roush, M. L., & Radosevich, S. R. (1985). Relationships Between Growth and Competitiveness of Four Annual Weeds. *Journal of Applied Ecology*, 22(3), 895-905. JSTOR. https://doi.org/10.2307/2403238

Roussel, S., &Triolo, J., (2016). *Bilan des opérations de lutte contre les plantes exotiques envahissantes menées par l'Office National des Forêts entre 2004 et 2013*. Direction régionale de l'ONF Réunion, Saint-Denis, La Réunion.

Saidou, W., Adama, K., Balé, B., & Amadou, T. (2015). Comparative in vitro anthelmintic activity on Hæmonchus contortus of two normal forages (casse obtusifolia and Piliostigma reticulatum). Extracts used in Burkina Faso, *international journal of agriculture and forestry* 5 (2), 146 -150.

Salmi M.,(2015). *Morpho-physiological and biochemical characterization of a few F2 generations of durum wheat (Triticum durum Desf.) under semi-arid conditions*. Magister enagronomie. Université Ferhat Abbas Sétif 1 Faculté des Sciences de la Nature et de la Vie.124p

Sanfo, A., Nouhoun, Z., & Kulo, E. (2020). Analysis of agropastoralists' preferences for forage production and conservation based on improved dual-purpose crop varieties in two agro-ecological zones in Burkina Faso. *Journal of Animal & Plant Sciences*, 46(3), 8318-8335.

Santiago L S., Kitajima K., Wright S J., & Mulkey S., (2004). Coordinated changes in photosynthesis, water relations and leaf nutritional traits of canopy trees along a precipitation gradient in lowland tropical forest. *Oecologia,* 139: 495-502.

Sare S. (2004). *Potentialités fourragères et effets de l'élevage extensif sur la diversité végétale dans la réserve de Biosphère de la mare aux hippopotames (ouest Burkinabé)*. Mémoire d'ingéniorat, IDR, Bobo-Diolasso, 92p.

Sarkar, E., Nath Chatterjee, S., & Chakraborty, P. (2012). Effect of cassia tora allelopathy on seed germination and growth of mustard. *Turkish Journal of Botany*, 36, 488 -494.

Savić, A., Oveisi, M., Božić, D., & Pavlović, D., (2021). Competition between Ambrosia artemisiifolia and Ambrosia trifida: Is there a threat of a stronger competitor?. *Weed Research*, 61(4), 298-306. https://doi.org/10.1111/wre.12479

Sawadogo I. (2011). *Forage resources and herders' representations, evolution of pastoral practices in a protected area context: Case du terroir de Kotchari à la périphérie de la réserve de biosphère du W au Burkina Faso. Environment and Society.* Doctoral thesis, Museum national d'histoire naturelle - MNHN PARIS, 2011. French. <tel-00708327>, 336p.

Schmidt, M., Kreft, H., Thiombiano, A., & Zizka, G. (2005). Herbarium collections and field data-based plant diversity maps for Burkina Faso. *Diversity and Distributions*, 11(6), 509-516.

Schmidt, M., Thiombiano, A., Zizka, A., König, K., Brunken, U., & Zizka, G. (2011). Plant functional trait models in biogeography of West African grasses (Poaceae). *African. Journal of Ecology*. 49, 490 -500.

Shackleton, S. E., & Shackleton, R. T. (2018). Local knowledge regarding ecosystem services and disservices from invasive alien plants in the arid Kalahari, South Africa. *Journal of arid environments*, 159, 22-33.

Shackleton, R. T., Richardson, D. M., Shackleton, C. M., Bennett, B., Crowley, S. L., Dehnen-Schmutz, K., Estévez, R. A., Fischer, A., Kueffer, C., & Kull, C. A. (2019). Explaining people's perceptions of invasive alien species: A conceptual framework. *Journal of Environmental Management*, 229, 10-26.

Seebens, H., Blackburn, T.M., Dyer, E.E., Genovesi, P., Hulme, P.E., Jeschke, J.M., Pagad, S., Pyšek, P., Winter, M., Arianoutsou, M., Bacher, S., Blasius, B., Brundu, G., Capinha, C., Celesti-Grapow, L., Dawson, W., Dullinger, S., Fuentes, N., Jäger, H., Kartesz, J., Kenis, M., Kreft, H., Kühn, I., Lenzner, B., Liebhold, A., Mosena, A., Moser, D., Nishino, M., Pearman, D., Pergl, J., Rabitsch, W., Rojas-Sandoval, J., Roques, A., Rorke, S., Rossinelli, S., Roy, H.E., Scalera, R., Schindler, S., Štajerová, K., Tokarska-Guzik, B., van Kleunen, M., Walker, K., Weigelt, P., Yamanaka, T., & Essl, F., (2017). No saturation in the accumulation of alien species worldwide. *Nature Communications* 8, 14435. https://doi.org/10.1038/ncomms14435

Senseman, S. A. & L. R. Oliver (1993). Flowering patterns, seed production, and somatic polymorphism of three weed species. *Weed Sci.* 41: 418-425.

Shackleton, S. E., & Shackleton, R. T. (2018). Local knowledge regarding ecosystem services and disservices from invasive alien plants in the arid Kalahari, South Africa. *Journal of arid environments,* 159, 22-33.

Shackleton, R. T., Larson, B. M. H., Novoa, A., Richardson, D. M., & Kull, C. A. (2019). The human and social dimensions of invasion science and management. *Journal of Environmental Management,* 229, 1-9. https://doi.org/10.1016/j.jenvman.2018.08.041

Schmidt, M., Thiombiano, A., Zizka, A., König, K., Brunken, U., & Zizka, G. (2011). Patterns of plant functional traits in the biogeography of West African grasses (Poaceae). *African Journal of Ecology* 49, 490-500.

Sheley, R. L., & Krueger-Betterave forage, J. (2003). Principles for reconstituting invasive plant-infested shotgrass. *La Science D'Mauvaise herbe*, 51, 260 -265.

Sieza, Y., Gomgnimbou, A. P., Serme, I., & Belem, A. (2019). Study of climatic variability on the dynamics of land occupation and use for agro-pastoral purposes in the South Sudanian zone of Burkina Faso. *International Journal of Biological and Chemical Sciences*, 13(4), 1980-1994.

Sinsin B (2001) Form de vie et diversité spécifique des associations de forets claires du Nord du Benin. *Syst. Georg. Pl.* 71 :873-888.

Sirima k. Maximin, (2022*). Sahel states facing the test of the terrorist hydra from 2015 to 2022.* Review University without Border for the Open Society (RUFSO).

Slama A., Salem M B., Naceur M B., & Zid E., (2005). Cereals in Tunisia: production, drought effect and resistance mechanisms. *Sécheresse* vol. 16, n° 3, September 2005, 5p.

Smit, G. N. (2004). A thinning tree approach to structuring southern African savannas for long-term bush encroachment restoration. *Journal of environmental management*, 71, 179 - 191.

Solbrig, O. T., Curtis, W. F., Kincaid, D. T., & Newell, S. J. (1988). Studies on the Population Biology of the Genus Viola. VI. The Demography of V. Fimbriatula and V. Lanceolata. *Journal of Ecology*, 76(2), 301-319. JSTOR. https://doi.org/10.2307/2260595

Solomon, G. (2015). Community perception on rangeland degradation: A case study in two differently settled areas of northern Ethiopia. *Journal of Agricultural Research and Development*, 5(1), 101-107.

Solomon G, &Yayneshet T (2014) Rangeland vegetation responses to settlement in the semi-arid rangelands of northern Ethiopia. *Scholarly Journal of Agriculture Science* 4:587-595

Some N. A., Traoré K., Traoré O., & Tassembedo M. (2007). Potential of artificial fallows with Andropogon spp. in improving the chemical and biological properties of soils in the Sudanian zone (Burkina Faso). *Biotechnology, Agronomy and Society and Environment*, 11(3), 351-357.

Somers, B., & Asner, P. G. (2013). Hyperspectral Multi-temporal analysis and mixing mode selection for invasive species tracer in rainforests. *Remote sensing of the environment,* 136, 14 -27.

Sop, T. K., Oldeland, J., Bognounou, F., Schmiedel, U., & Thiombiano, A. (2012). Ethnobotanical knowledge and valuation of woody plants species: A comparative analysis of three ethnic groups from the sub-Sahel of Burkina Faso. *Environment, Development and Sustainability*, 14(5), 627-649. https://doi.org/10.1007/s10668-012-9345-9

Souley, M. S., Kiari, S. A., Morou, B., Synnevag, G., & Aune, J. B. (2020). Peasant perception of sida cordifolia and socioeconomic factors influencing adoption of its compost in western and central Niger.

SP/CONEDD: Secrétariat Permanent du Conseil National pour l'Environnement et le Développement Durable (2010). Burkina Faso's fourth national report on biological diversity. Ouagadougou, 119p.

SP/CONEDD: Permanent Secretariat of the National Council for the Environment and Sustainable Development (2014). Burkina Faso's fifth national report on biological diversity. Ouagadougou, 104p.

Sy, A., Grouzis, M., Danthu, P. (2001). Short communication: Seed germination of seven Sahelian legume species. *Journal of Arid Environments*, 49, 875 -882.

Synowiec, A., Jop, B., Domaradzki, K., Podsiadło, C., Gawęda, D., Wacławowicz, R., & Praczyk, T. (2021). Environmental factors effects on winter wheat competition with herbicide-resistant or susceptible silky bentgrass (Apera spica-venti L.) in Poland. *Agronomy*, 11(5), 871.

Tassin J., Thompson K., Carroll S.P., Thomas C.D., & (2017). Determining whether the impacts of introduced species are negative cannot be based solely on science: a response to Russell and Blackburn. *Trends in Ecology & Evolution,* 32 (4): : e00001230-231

Teem, D. H., Hoveland, C. S., & Buchanan, G. A. (1980). Sicklepod (Cassia obtusifolia) and coffee senna (Cassia occidentalis): geographic distribution, germination, and emergence. *Weed Science*, 28(1), 68-71.

Teem, D. H., Hoveland C. S., & Buchanan G. A. (1974). Primary root elongation of three weed species. *Weed Science.* 22:47-50.

Thevenot J. (coord.) (2013). Synthesis and reflections on definitions related to biological invasions. Preamble to the actions of the national strategy on invasive alien species (IAS) having a negative impact on biodiversity. Report SPN 2013/15, Muséum National d'Histoire Naturelle, Service du Patrimoine Naturel, Paris, 31 p.

Thiombiano D., (2008). *Etude de trois espèces à tendance prolifique et envahissante dans la province du Boulkiemdé : Essai de valorisation (Burkina Faso).* Mémoire d'Ingénieur des Eaux et Forêts, IDR/UPB, 127P.

Thiombiano N., Ouedraogo L.R., Belem M., & Guinko S., (2009). Evolutionary dynamics and impact of an invasive plant in Burkina Faso: Hyptis suaveolens (l.) Poit. Ann Univ Lomé Togo *Série* Sciences, 1897-115

Thiombiano A. & Kampmann D. (2010*). Atlas de la Biodiversité de l'Afrique de l'Ouest, Tome II : Burkina Faso.* Ouagadougou & Frankfurt/Main. Ouagadougou and Frankfurt. 592p.

Thiombiano, A., Schmidt M., Dressler, S., Hahn, K., & Zizka, G. (2012). *Catalog of the vascular plants of Burkina Faso.* Boissiera 65, 391p.

Tordoni, E., Petruzzellis, F., Nardini, A., Savi, T., & Bacaro, G. (2019). Make it simpler: Alien species decrease functional diversity of coastal plant communities. *Journal of Vegetation Science*, 30(3), 498-509.

Traoré M., (1996). *Utilisation des éléments nutritifs par une graminée pérenne : Andropogon gayanus*, Thèse pour obtenir le titre de Docteur de Spécialité, Option : Production Fourragère à l'Institut Supérieur de Formation et de Recherche Appliquée, Institut d'Economie Rurale (IER), 165p.

Traore L., Ouédraogo I., Ouedraogo A., & Thiombiano A., (2011). Perceptions, uses and vulnerability of woody plant resources in Southwest Burkina Faso. *International Journal of Biological and Chemical Sciences*, 5(1), 258-278. https://doi.org/10.4314/ijbcs.v5i1.68103.

Traoré, S., Thiombiano, L., Bationo, B. A., Kögel-Knabner, I., & Wiesmeier, M. (2020). Organic carbon fractional distribution and saturation in tropical soils of West African savannas with contrasting mineral composition. *Catena*, 190, 104550.

Tresch and Fliessbach (2017). *Study of decomposition using tea bags. Fertil Crop Technical Note*. Download from www.fertilcrop.net, 4p.

Triplet, P. (2012). *Manuel de gestion des aires protégées d'Afrique francophone*. Awely, Paris, 2009. <hal-00669157>. pp.1215.

Tungate KD, Susko DJ, & Rufty TW (2002) Reproduction and offspring competitiveness of *Senna obtusifolia* are influenced by nutrient availability.*New Phytol* 154:661-669

Tye A, Soria MC, & Gardener MR (2003) A strategy for Galapagos weeds. In: Veitch CR, Clout MN (eds) Turning the tide: the eradication of invasive species. *Proceedings of the international conference on eradication of island invasive*. IUCN, Gland, pp 336-341

IUCN, (2004). *Prevention and management of invasive alien species: Implementing Cooperation in West Africa*. Proceedings of a Regional Workshop ,117p.
IUCN-Burkina Faso (2015). *Assessment of the general state of pastoral resources in Burkina Faso Ouagadougou, Burkina Faso*: IUCN. 118 p.

IUCN. (2017). *Study on the realization and documentation of the reference situation of the communal forest of Dori and its zone of influence*. The northern forest of the first Cordon Dunaire of Burkina Faso. Burkina Faso IUCN.79 p.

Upadhyaya SK, & Singh V, (1986). Phytochemical evaluation of Cassia obtusifolia L. and Cassia tora L. *Proceedings of the Indian Academy of Sciences, Plant Sciences*, 96(4):321-326.

Vallé C., Blodeau G. and Lanaudière C. J., (1999). *Multi-cell culture techniques*. Les presses de l'université Laval. 394p

Valéry L., (2006). *Approche systémique de l'impact d'une espèce invasive: le cas d'une espèce indigène dans un milieu en voie d'eutrophisation*. Thèse de doctorat en Écologie, Paris, Muséum national d'histoire naturelle, Paris, France, 276 p.

Valery L., Fritz H., Lefeuvre J.-C. & Simberloff D. (2008). - In search of a real definition of the biological invasion phenomenon itself. *Biological Invasions* 10: 1345-1351.

Vankat, J. L., & Graham, R. D. (2002). Landscape invasion by exhotic species: Gützweiler, K. J. (ED), Applying landscape ecology in biological conservation. *Biology Conservation*. Springer, New York, 171-191.

Vila, M., & Ibáñez, I., (2011). Plant invasions in the landscape. *Landscape ecology* 26, 461-472.

Voss, K. A., & Bernecke, L. H. (1991). Toxiological and hematological effects of sicklepod seeds (cassia obtusifolia) in Sprague-Dawley rats: a subchronic feeding study. *Toxicon*, 29: 1329-1336.

Waddell, E. H., Banin, L. F., Fleiss, S., & Hill, J. K., (2020). Land-use change and propagule pressure promote plant invasions in tropical rainforest remnants. *Landscape Ecology*, *35*(9), 1891-1906. https://doi.org/10.1007/s10980-020-01067-9

Wagh, V. V., & Jain, A. K. (2018). Status of ethnobotanical invasive plants in western Madhya Pradesh, India. *South African Journal of Botany*, 114, 171-180. https://doi.org/10.1016/j.sajb.2017.11.008

Walker, E. R., & Oliver, L. R. (2008). Translocation and absorption of glyphosate in flowering sicklepod (*Senna obtusifolia*). *Weed Science*, v. 56, n. 3, p. 338-343,

Walsh, J. R., Carpenter, S. R., & Vander Zanden, M. J. (2016). Invasive species trigger a massive loss of ecosystem services through a trophic cascade. *Proceedings of the National Academy of Sciences*, 113(15), 4081-4085.

Wang, J., Fan, X., Fan, J., Zhang, C., & Xia, F. (2012). Effect of aboveground competition on biomass partitioning of understory Korean pine (Pinus koraiensis). Shengtai Xuebao/Acta *Ecologica Sinica*, 32(8), 2447-2457.

Wardle, D. A., Bardgett, R. D., Callaway, R. M., & Putten, W. H. V. der. (2011). Terrestrial Ecosystem Responses to Species Gains and Losses. *Science*, 332(6035), 1273-1277. https://doi.org/10.1126/science.1197479

Webster, T. M., & Macdonald, G. E. (2001). A Survey of Weeds in Various Crops in Georgia. *Weed Technology*, 15(4), 771-790. https://doi.org/10.1614/0890-037X(2001)015[0771:Asowin]2.0.CO;2

Weigelt, A., & Jolliffe, P. (2003). Indices of plant competition. *Journal of Ecology*, 91(5), 707-720. https://doi.org/10.1046/j.1365-2745.2003.00805.x

Wenda-Piesik, A., Synowiec, A., Marcinkowska, K., Wrzesińska, B., Podsiadło, C., Domaradzki, K., .& Kwiecińska-Poppe, E. (2022). Intra-and interspecies competition of blackgrass and wheat in the context of herbicidal resistance and environmental conditions in Poland. *Scientific Reports*, 12(1), 1-16

White F (1986). *The vegetation of Africa. Memoir accompanying the vegetation map of Africa*. Unesco/AETFAT /UNSO, OROSTOM.

Wiegand, K., Saltz, D., & Salle, D. (2006). *A piece-dynamics approach to savanna dynamics and woody* encroachment-plant *insights* from an arid savanna. Perspectives in phytoecology, evolution and systematics, 7, 229 -242.

Wickert, K. L., O'Neal, E. S., Davis, D. D., & Kasson, M. T. (2017). Seed Production, Viability, and Reproductive Limits of the Invasive Ailanthus altissima (Tree-of-Heaven) within Invaded Environments. *Forests*, 8(7), 226. https://doi.org/10.3390/f8070226

Wijewardana, C., Alsajri, F. A., Irby, J. T., Krutz, L. J., Golden, B., Henry, W. B., Gao, W., & Reddy, K. R. (2019). Physiological assessment of water deficit in soybean using midday leaf water potential and spectral features. *Journal of Plant Interactions*, 14(1), 533-543.

Willey, R. W., & Rao, M. R. (1980). A Competitive Ratio for Quantifying Competition Between Intercrops†. *Experimental Agriculture*, 16(2), 117-125. https://doi.org/10.1017/S0014479700010802

Williams, A. P., & Chasse, E. De J. R. (2002). The assessment of spurge hardwood cover from hyperspectral figurative language employing mixture tuned matched filtering. *Remote sensing of the environment*, 82, 446 - 456.

Williamson M. H. & Fitter A. (1996). The characters of successful invaders. *Biological Conservation* 78 (1996) 163-170. www.teatime4science.org

Wittig, R., König, K., Schmidt, M., & Szarzynski, J. (2007).A study of climate change and anthropogenic impacts from West Africa, Environmental Science and Pollution Research - International, *Weed Science.* 14 (3), 182 -189.

Wright, S. R., Raper, C. D. Jr., & Ruffy, T. W. Jr. 1999. Comparative responses of soybean (*Glycine max*), sicklepod (*Senna obtusifolia*), and Palmer amaranth (*Amaranthus palmeri*) to root zone and aerial temperatures. *Weed Science.* 47:167-174.

Wu, J., Wang, J., Hui, W., & Zhao, F., (2022). Physiology of Plant Responses to Water Stress and Related Genes: *A Review. Forests,* 13(2), 324. https://doi.org/10.3390/f13020324

Yameogo G., Kiema A., Yelemou B., & Ouedraogo L. (2013). Characteristics of herbaceous forage resources of natural pastures in the Vipalogo terroir (Burkina Faso). *International Journal of Biology and Chemical* Sciences 7(5): 2078-2091.*

Yang, X., Zhang, W., & He, Q. (2019). Effects of intraspecific competition on growth, architecture and biomass allocation of Quercus Liaotungensis. *Journal of Plant Interactions*, 14(1), 284-294. https://doi.org/10.1080/17429145.2019.1629656

Yang, X., Lu, M., Wang, Y., &Wang, Y., (2021). Response mechanism of plants to drought stress. *Horticulturae,* 7(3), 50. https://doi.org/10.3390/horticulturae7030050

Yohann.S. (2015). *Practical guide for early detection and rapid response to invasive alien species in French overseas collectivities.* General principles, guidelines and implementation options. IUCN French Committee, Paris, France, 75 p. ISBN : 978-2-918105-47-3

Yu, H., Shen, N., Yu, S., Yu, D., & Liu, C. (2018). Responses of the native species Sparganium angustifolium and the invasive species Egeria densa to warming and interspecific competition. *PloS one*, 13(6), e0199478.

Zampaligre, N., Kagambega, F. W., Zida, D., Traore, C. T., & Sawadogo, L. (2020). Floristic composition, diversity and structures of grazed savanna woody plants: the case of the Dinderesso sylvopastoral zone, Burkina Faso. *Afrique Science*, 16(1), 78-91.

Zand, E., & Beckie, H. J. (2002). Competitive ability of hybrid and open-pollinated canola (Brassica napus) with wild oat (Avena fatua). *Canadian Journal of Plant Science*, 82(2), 473-480. https://doi.org/10.4141/P01-149

Zerbo I. (2011). *Spatio-temporal dynamics and productivity of Sudano-Sahelian and Sudanian pastures in Burkina Faso*. DEA thesis in Applied Biological Sciences from the University of Ouagadougou. 61p.

Zerbo, I., Bernhardt-Römermann, M., Ouédraogo, O., Hahn, K., & Thiombiano, A. (2016). Effects of climate and land use of herbaceous species richness and vegetation composition in West African savanna ecosystems. *Journal of Botany*,2016 :1-11.

Zerbo I.,(2018). *Functional diversity and responses of savanna herbaceous vegetation to variations in environmental factors in Burkina Faso (West Africa)*. These de doctorat unique en Sciences Biologiques appliquées de l'Université Ouaga I Pr Joseph Ki-Zerbo. 147p.

Zhang, G., Yang, Z., & Dong, S. (2011). Interspecific competitiveness affects the total biomass yield in an alfalfa and corn intercropping system. *Field Crops Research*, 124(1), 66-73. https://doi.org/10.1016/j.fcr.2011.06.006

Zhou, W., Cheng, X., Wu, R., Han, H., Kang, F., Zhu, J., & Tian, P. (2018). Effect of intraspecific competition on biomass partitioning of Larix principis-rupprechtii. *Journal of Plant Interactions,* 13(1), 1-8.

Zizka, A., Thiombiano, A., Dressler, S., Nacoulma, B. M., Ouédraogo, A., Ouédraogo, I., ... & Schmidt, M. (2015). Traditional plant use in Burkina Faso (West Africa): a national-scale analysis with focus on traditional medicine. *Journal of Ethnobiology and Ethnomedicine*, *11*(1), 1-10. https://link.springer.com/article/10.1186/1746-4269-11-9

Zuur, A. F., Ieno, E. N., Walker, N., Saveliev, A. A., Smith, G. M., Zuur, A. F., & Smith, G. M. (2009). Zero-truncated and zero-inflated models for count data. *Mixed effects models and extensions in ecology with R*, 261-293

APPENDICES

Appendix 1Ethnobotanical survey form

I.   Identification

Surveyor.........Date.........Report number

Commune.........Localité..............

Identity of respondent

Survey (e) .... Gender...........(1=woman; 2=man)

Age:.......Ethnicity..............Occupation.....Level of education..........

(1=not educated ;2=primary ;3=secondary (specify class) ;4=higher (specify level)

Respondent contact..

II. Ecological knowledge of the survey

Do you know this species (present sample taken)? yes no

What do you call it in your language?

Or is the species abundant in your area?

Around habitas (houses)....in fields......in fallow land.......in forests.....in lowlands.......in glacis......other locations (please specify)

When did the species first appear in your area? (number of years)

III-Factors favoring the species' expansion

What factors do you think encourage the expansion of *Senna obtusifolia*? (Let the survey quote)

a-anthropogenic factors (man, woman, animals)..................

b-Climatic factors (rain, wind, temperature, etc.)...................

c-Edaphic factors (soil types)....................................

d-Other (please specify).......................................

IV-Performance and impact of *Senna obtusifolia* on herbaceous forage and soil

a-Does Senna *obtusifolia* grow in large or small quantities from the first rains?

....

b- Does the presence of *Senna obtusifolia* decrease or increase each year?

2- Impact of *Senna obtusifolia* on herbaceous forage

a-When *Senna obtusifolia* becomes established, does the density of other herbaceous species decrease or increase? Why or why not?

....

b-Is the herbaceous forage in areas heavily invaded by Senna frequented by cattle or other domestic animals? Why or why not?

....

c-Do you know of other invasive species that are not palatable? Examples...

3- Impact of *Senna obtusifolia* on the soil

Does *Senna* impoverish or enrich the soil? And why?

....

IV- Actions and strategies to mitigate the impact of land invasion by *S. obtusifolia*

1-Do you know how to manage the species? yes no

If so, which ones?

....

2-What management method do you recommend?

a-Pulverize b-harvest **c-harvest** leaves **d-harvest** flowers **e-harvest** stems **f-harvest** as fodder g-other

What strategies have you put in place to obtain forage?

V-Different uses of the species by populations

1- Do you use the species? If yes (table)

Type of use	Organs used	Collection period	Species maturity	Degree of use	How to use	Method of administration
Food						
Medicinal						
Construction						
Forage						
Crafts						
Trade						
Wood and energy						
Other uses						

Type of use: 1= food; 2= medicinal; 3= construction; 4= fodder; 5= handicrafts; 6= wood and energy; 7= other uses.
Organs used: 1= Root; 2= Bark; 3= Leaves; 4= Fruit; 5= Seed; 6= Flower.
Collection period: 1= Dry season; 2= Rainy season.
Species maturity: 1=juvenile; 2= adult.
Degree of use: 1= not used; 2= lightly used; 3= moderately used; 4= heavily used.
Instructions for use :1=boil , 2=let dry,
Method of administration: 1=oral,2=decoction
2- What are the traditional prohibitions associated with this species?
....
....
....
3- Is *the* .... *species* used in traditional rites?.......................................
....

Appendix 2Phytosociological data collection sheet for plots

Plot number: .... Location: .... Date : .... GPS :...............
Density of *S. obtusifolia*: ....biomass: .... Biomass of other species:
....
Plot: ....

Climate zone	Land use	Location	Topography	Invasion level of *S. obtusifolia*
• Sahelia n zone Sudanian zone	• Grazin g • Fallow land	• 1 • 2 • 3 • 4 • 5	• Glazin g • Bafond	• T 0% • A 5-35% • B 35-65%

Species name				Height (H)	• C 65-100%
					Recovery rate n (%)

Appendix 3List of species inventoried in the two climatic zones (Sahel; Sudan)

	Species	**Family**	**TB**	**TP**	**VP**
1	*Acalypha ciliata* Forsk.	Euphorbiaceae	Tea	PAL	FVP
2	*Acanthospermum hispidum* DC.	Asteraceae	Pha	PAN	SVF
3	*Achyranthes aspera* L.	Amaranthaceae	Cha	PAN	PLEASE
4	*Acmella uliginosa* (Sw.) Cass.	Asteraceae	Pha	PAN	PLEASE
5	*Acroceras zizanoides* Kunth	Poaceae	Cha	AT	MVP
6	*Aeschynomene indica* Linn.	Fabaceae	Tea	PAN	MVP
7	*Ageratum conyzoides* L.	Asteraceae	Tea	PAN	PLEASE
8	*Alternanthera nodiflora* R. Br.	Amaranthaceae	Cha	AT	PLEASE
9	*Alternanthera sessilis* (L.) DC.	Amaranthaceae	Cha	AT	PLEASE
10	*Alysicarpus ovalifolius* (S. et Th.) Leon	Fabaceae	Tea	PAN	BVP
11	*Alysicarpus rugosus* (Willd.) DC.	Fabaceae	Tea	PAN	BVP
12	*Amaranthus graecizans* L.	Amaranthaceae	Tea	PAN	MVP
13	*Amaranthus spinosus* L.	Amaranthaceae	Tea	COS	MVP
14	*Amaranthus viridis* L.	Amaranthaceae	Tea	COS	MVP
15	*Andropogon gayanus* Kunth	Poaceae	Hem	PAN	BVP
16	*Aneilema setiferum* A.Chev.	Commelinaceae	Cha	AT	FVP
17	*Annona senegalensis* Pers.	Annonaceae	Pha	AM	FVP
18	*Anogeissus leiocarpa* (DC) Guill. et Perr.	Combretaceae	Pha	AT	FVP
19	*Aristida adscensionis* L.	Poaceae	Tea	PAN	MVP
20	*Aristida kerstingii* Pilg.	Poaceae	Tea	PAN	FVP
21	*Aspilia bussei* O. Hoffm. & Muschler	Asteraceae	Tea	AT	MVP
22	*Aspilia helianthoides* (Schum. & Thonn.)	Asteraceae	Tea	AT	MVP
23	*Azadirachta indica* A. Juss.	Meliaceae	Pha	PAN	PLEASE

	Species	Family	TB	TP	VP
24	*Balanites aegyptiaea* (L.) Del.	Zygophyllaceae	Pha	PAN	BVP
25	*Bidens engleri* O. E. Schulz	Asteraceae	Tea	AT	PLE ASE
26	*Biophytum petersianum* Hotzsch.	Oxalidaceae	Tea	PAL	PLE ASE
27	*Blepharis maderaspatensis* (L.) Roth	Acanthaceae	Tea	PAN	FVP
28	*Boerhavia diffusa* L.	Nyctagynaceae	Tea	PAN	PLE ASE
29	*Bombax costatum* Pell. et Vuill.	Malvaceae	Pha	AT	BVP
30	*Borassus akeassii* Bayton, Ouédr. & Guinko	Arecaceae	Pha	PAN	FVP
31	*Brachiaria lata* (Schumach.) C. E.Hubbard	Poaceae	Tea	PAN	BVP
32	*Brachiaria stigmatisata* (Metz) Stapf	Poaceae	Tea	AT	MVP
33	*Brachiaria villosa* (Lam.) A.Camus	Poaceae	Tea	PAL	MVP
34	*Buchnera hispida* Buch. Ham. Ex. Benth	Orobanchaceae	Tea	PAN	FVP
35	*Bulbostylis filamentosa* (Vahl) C.B. Cl.	Cyperaceae	Tea	AT	PLE ASE
36	*Bulbostylis hispidula* (Vahl) R.W.Haines	Cyperaceae	Tea	AT	PLE ASE
37	*Calopogonium mucunoides* Desv.	Fabaceae	Pha	PAN	FVP
38	*Carissa edulis* Vahl	Apocynaceae	Pha	PAN	MVP
39	*Cassia absus* L.	Fabaceae	Tea	PAN	FVP
40	*Cassia nigricans* Vahl.	**Fabaceae**	Tea	PAN	FVP
41	*Cassia sieberiana* DC.	**Fabaceae**	Pha	AT	FVP
42	*Celosia trigyna* L.	Amaranthaceae	Tea	PAN	FVP
43	*Cenchrus biflorus* Roxb.	Poaceae	Tea	PAN	BVP
44	*Chamaecrista mimosoides* (L.) Greene	Fabaceae	Tea	PAL	MVP
45	*Chloris gayana* Kunth	poaceae	Tea	PAN	BVP
46	*Chloris pilosa* Schum.	poaceae	Tea	PAN	MVP
47	*Chlorophytum blepharophyllum* Schweinf. ex Baker	Antheriaceae	Geo	SZ	FVP
48	*Chrysanthellum americanum* (L.) Vatke	Asteraceae	Tea	PAN	PLE ASE
49	*Citrullus colocynthis* (L.) Schrader	Cucurbitaceae	Pha	AT	FVP
50	*Cleome viscosa* L.	Capparaceae	Tea	PAN	FVP
51	*Cochlospermum planchonii* Hook. f.	Bixaceae	Cha	AT	MVP

	Species	Family	TB	TP	VP
52	*Combretum aculeatum* Vent.	Combretaceae	Pha	AT	PLE ASE
53	*Combretum collinum* Fresen.	Combretaceae	Pha	AT	PLE ASE
54	*Combretum glutinosum* Perr. ex DC.	Combretaceae	Pha	AT	PLE ASE
55	*Combretum micrantum* G. Don	Combretaceae	Pha	AT	PLE ASE
56	*Combretum molle* R. Br. ex G. Don	Combretaceae	Pha	AT	PLE ASE
57	*Commelina benghalensis* L.	Commelinaceae	Cha	PAN	MVP
58	*Commelina diffusa* Burm.f.	Commelinaceae	Cha	PAN	MVP
59	*Commelina forskalaei* Vahl.	Commelinaceae	Cha	PAN	MVP
60	*Corchorus olitorius* L.	Malvaceae	Tea	PAN	MVP
61	*Corchorus tridens* L.	Malvaceae	Tea	PAN	MVP
62	*Cordia myxa* L.	Boraginaceae	Pha	PAN	PLE ASE
63	*Crotalaria goreensis* Guill.& Perr.	Fabaceae	Cha	PAN	PLE ASE
64	*Crotalaria macrocalyx* Benth.	Fabaceae	Cha	AT	FVP
65	*Crotalaria retusa* L.	Fabaceae	Tea	PAN	FVP
66	*Cucumus melo* L.	Cucurbitaceae	Tea	PAN	MVP
67	*Cyanotis lanata* Benth.	Commelinaceae	Tea	AT	BVP
68	*Cyperus difformis* Linn.	Cyperaceae	Tea	PAN	FVP
69	*Cyperus dilatatus* Schumach. & Thonn.	Cyperaceae	Geo	AT	FVP
70	*Cyperus esculentus* Linn.	Cyperaceae	Geo	PAN	PLE ASE
71	*Cyperus haspan* Linn.	Cyperaceae	Tea	PAN	FVP
72	*Cyperus iria* Linn.	Cyperaceae	Tea	PAN	FVP
73	*Cyperus pustulatus* Vahl	Cyperaceae	Tea	AT	FVP
74	*Cyperus rotundus* L.	Cyperaceae	Geo	PAN	FVP
75	*Cyperus sphacelatus* Rottb.	Cyperaceae	Hem	PAN	FVP
76	*Dactyloctenium aegyptium* (Linn) Wild.	Poaceae	Tea	PAL	BVP
77	*Daniellia oliveri* (R.) Hutch. & Dalz.	Fabaceae	Pha	AT	PLE ASE
78	*Desmodium gangeticum* (L.) DC	Fabaceae	Tea	PAL	FVP
79	*Desmodium hirtum Guill* & Perr.	Fabaceae	Tea	AT	MVP
80	*Detarium microcarpum* G. & Perr.	Fabaceae	Pha	AT	PLE ASE

	Species	Family	TB	TP	VP
81	*Dichrostachys cinerea* L.Wight & Arn.	Fabaceae	Pha	AM	BVP
82	*Digitaria horizontalis* Willd.	poaceae	Tea	PAN	MVP
83	*Dinebra retroflexa* (Vahl) Panz.	poaceae	Tea	PAN	BVP
84	*Dioscorea alata* L.	Dioscoreaceae	Geo	PAN	FVP
85	*Dioscorea dumetorum* (Kunth) Pax	Dioscoreaceae	Geo	AT	MVP
86	*Diospyros mespiliformis* Hochst. ex A. Rich.	Ebenaceae	Pha	AT	BVP
87	*Echinochloa colona* (L.) Link	Poaceae	Tea	PAN	MVP
88	*Echinochloa stagnina* (Retz.) P.Beauv.	Poaceae	Tea	PAN	MVP
89	*Eleusine indica* (L.) Gaertn.	Poaceae	Tea	PAN	FVP
90	*Elionurus elegans* Kunth	Poaceae	Tea	AT	MVP
91	*Entada africana* Guill. et Perr.	Fabaceae	Pha	AM	BVP
92	*Eragrostis atrovirens* (Desf.) Trin. Ex Steud.	Poaceae	Cha	AT	MVP
93	*Eragrostis ciliaris* (L.) R. Br.	Poaceae	Tea	PAN	MVP
94	*Eragrostis tenella* (L.) Roem. & Schult.	Poaceae	Tea	AT	MVP
95	*Eragrostis tremula* Hochst.	Poaceae	Tea	AT	MVP
96	*Eriosema psoraleioides* (Lam.) G. Don.	Fabaceae	Cha	AT	FVP
97	*Erythrina senegalensis* DC.	Fabaceae	Pha	AT	FVP
98	*Euclasta condylotricha* (Steud) Stapf.	Poaceae	Tea	PAN	MVP
99	*Euphorbia heterophylla* L.	Euphorbiaceae	Tea	PAN	MVP
100	*Euphorbia hirta* L.	Euphorbiaceae	Tea	PAN	MVP
101	*Evolvolus alsinoides* (Linn.)	Convolvulaceae	Tea	AT	FVP
102	*Faidherbia albida* (Delile) A.Chev.	Fabaceae	Pha	AT	BVP
103	*Feretia apodanthera* Del.	Rubiaceae	Pha	AM	MVP
104	*Ficus ingens* (Miq.) Miq.	Moraceae	Pha	PAN	PLE ASE
105	*Ficus sycomorus* L.	Moraceae	Pha	AM	PLE ASE
106	*Fimbristylis ferurginea* (Linn.) Vahl	Cyperaceae	Pha	PAN	FVP
107	*Flueggea virosa* (Roxb. ex Willd.) Voigt	Phyllanthaceae	Pha	PAN	FVP
108	*Gardenia erubescens* Staf et Hutch	Rubiaceae	Pha	AT	MVP
109	*Gardenia ternifolia* Schumach. & Thonn.	Rubiaceae	Pha	AT	PLE ASE
110	*Gomphrena celosioides* Mart.	Amaranthaceae	Tea	PAN	FVP

	Species	Family	TB	TP	VP
111	*Grewia cissoides* Hutch. et Dalz.	Malvaceae	Cha	AT	MVP
112	*Grewia mollis* Juss.	Malvaceae	Pha	PAN	FVP
113	*Guiera senegalensis* J.F. Gmel.	Combretaceae	Cha	AT	FVP
114	*Gymnosporia senegalensis* (Lam.) Loes.	Celastraceae	Pha	PAN	BVP
115	*Hackelochloa granularis* (L.) Kuntze	Poaceae	Tea	PAN	MVP
116	*Hibiscus cannabinus* L.	Malvaceae	Tea	PAN	MVP
117	*Hygrophila auriculata* (Schumach.) Heine	Acanthaceae	Tea	PAN	FVP
118	*Hymenocardia acida* Tul.	Phyllanthaceae	Cha	AT	PLE ASE
119	*Hyphaene thebaica* (L.) Mart.	Arecaceae	Pha	PAN	PLE ASE
120	*Hyptis spicigera* Lam.	Lamiaceae	Tea	PAN	PLE ASE
121	*Hyptis suaveolens* Poit.	Lamiaceae	Tea	PAN	PLE ASE
122	*Indigofera bracteolata* DC.	Fabaceae	Cha	AT	MVP
123	*Indigofera heudelotii* Benth	Fabaceae	Tea	AT	FVP
124	*Indigofera hirsuta* L.	Fabaceae	Tea	PAN	MVP
125	*Indigofera mummulariifolia* (L.) Liv.	Fabaceae	Tea	AT	FVP
126	*Indigofera pilosa* Poir.	Fabaceae	Tea	AT	PLE ASE
127	*Indigofera senegalensis* Lam.	Fabaceae	Tea	AT	FVP
128	*Ipomoea aquatica* Forsk.	Convolvulaceae	Tea	PAN	FVP
129	*Ipomoea argentaurata* Hall. f.	Convolvulaceae	Tea	AT	MVP
130	*Ipomoea asarifolia* (Desr.) Roem. & Schult.	Convolvulaceae	Tea	AT	MVP
131	*Ipomoea coscinosperma* Hochst. ex Choisy	Convolvulaceae	Tea	AT	BVP
132	*Ipomoea eriocarpa* R. Br.	Convolvulaceae	Tea	PAL	BVP
133	*Justicia kotschyi* (Hochst.) Dandy	Acanthaceae	Cha	AT	MVP
134	*Khaya senegalensis* (Desr.) A. Juss.	Meliaceae	Pha	PAN	BVP
135	*Kyllinga erecta* Schumach.	Cyperaceae	Geo	PAN	PLE ASE
136	*Kyllinga pumila* Michx	Cyperaceae	Hem	PAN	PLE ASE
137	*Kyllinga squamulata* Thonn. ex Vahl	Cyperaceae	Tea	PAN	PLE ASE
138	*Lannea acida* A. Rich.	Anacardiaceae	Cha	AT	FVP

	Species	Family	TB	TP	VP
139	*Lannea microcarpa* Engl. & K.Krause	Anacardiaceae	Cha	AT	MVP
140	*Leptadenia hastata* (Pers.) Decne	Apocynaceae	Pha	AT	FVP
141	*Leucas martinicensis* (Jacq.) R. Br.	Lamiaceae	Tea	PAN	PLE ASE
142	*Lipocarpha chinensis* (Osbeck) Kern	Cyperaceae	Tea	PAN	PLE ASE
143	*Ludwigia abyssinica* A. Rich.	Onagraceae	Pha	AT	PLE ASE
144	*Ludwigia hyssopifolia* (G. Don.) Exell	Onagraceae	Tea	PAN	PLE ASE
145	*Maranthes polyandra* (Benth.) Prance	Chrysobalanaceae	Pha	AT	PLE ASE
146	*Mariscus cylindristachyus* Steud.	Cyperaceae	Hem	PAN	PLE ASE
147	*Mariscus flabelliformis* Kunth	Cyperaceae	Geo	PAN	PLE ASE
148	*Marsilea minuta* L.	Marsileaceae	Hel	PAN	PLE ASE
149	*Melochia corchorifolia* Linn.	Malvaceae	Tea	PAL	MVP
150	*Microchloa indica* (L.f.) P.Beauv.	Poaceae	Tea	PAN	PLE ASE
151	*Mitracarpus scaber* Zucc.	Rubiaceae	Tea	PAN	FVP
152	*Mollugo nudicaulis* Lam.	Molluginaceae	Tea	PAN	PLE ASE
153	*Murdannia simplex* (Vahl) Brenan	Commelinaceae	Pha	PAN	FVP
154	*Nelsonia canescens* (Lam.) Spreng.	Acanthaceae	Tea	PAN	FVP
155	*Ocimum canum* Sims	Lamiaceae	Tea	PAN	PLE ASE
156	*Oldenlandia corymbosa* Linn.	Rubiaceae	Tea	PAN	MVP
157	*Oryza longistaminata* Chev. & Roehr.	poaceae	Hel	AM	MVP
158	*Pandiaka heudelotii* (Moq.) Hook	Amaranthaceae	Tea	AT	BVP
159	*Panicum laetum* Kunth	Poaceae	Tea	AT	MVP
160	*Panicum laxum* Sw.	Poaceae	Tea	PAN	MVP
161	*Panicum pansum* Rendle	Poaceae	Tea	AT	FVP
162	*Panicum phragmitoides* Stapf	Poaceae	Hem	AT	FVP
163	*Parkia biglobosa* (Jacq.) Benth.	Fabaceae	Pha	AT	BVP
164	*Paspalum scrobiculatum* L.	Poaceae	Hem	PAN	BVP
165	*Pennisetum glaucum* (L.) R.Br	Poaceae	Tea	PAN	BVP
166	*Pennisetum pedicellatum* Trin.	Poaceae	Tea	PAN	BVP

	Species	Family	TB	TP	VP
167	*Pentanema indicum* (L.) Y.Ling	Asteraceae	Tea	PAL	PLE ASE
168	*Pericopsis laxiflora* (Benth.) Meeuwen	Fabaceae	Pha	AT	MVP
169	*Peristrophe bicalyculata* (Retz.) Nees	Acanthaceae	Tea	PAN	FVP
170	*Phaulopsis imbricata* (Forssk.) Sweet	Acanthaceae	Tea	AT	PLE ASE
171	*Phyllanthus amarus* Sch. Et Th.	Phyllanthaceae	Tea	PAN	FVP
172	*Physalis angulata* L.	Solanaceae	Tea	PAN	PLE ASE
173	*Piliostigma reticulatum* (Dc.) Hoschst	Fabaceae	Pha	AT	MVP
174	*Platostoma africanum* P. Beauv	Lamiaceae	Tea	PAN	PLE ASE
175	*Polycarpaea corymbosa* (L) Lam.	Caryophyllaceae	Tea	PAN	PLE ASE
176	*Polycarpaea linearifolia* Lam	Caryophyllaceae	Tea	AT	PLE ASE
177	*Polygala arenaria* Willds	Polygalaceae	Tea	AT	PLE ASE
178	*Polygala multiflora* Poiret	Polygalaceae	Tea	AT	PLE ASE
179	*Portulaca foliosa* Ker Gawl.	Portulacaceae	Tea	PAN	MVP
180	*Portulaca oleracea* L.	Portulacaceae	Tea	COS	MVP
181	*Prosopis africana* (G. et Perr.) Taub.	Fabaceae	Pha	AT	PLE ASE
182	*Pterocarpus erinaceus* Poir.	Fabaceae	Pha	AT	BVP
183	*Pycreus flavescens* (L.) P.Beauv. ex Rchb.	Cyperaceae	Hel	PAN	FVP
184	*Pycreus macrostachyos* (Lam.) J.Raynal	Cyperaceae	Tea	PAN	FVP
185	*Ramphicarpa fistulosa* (Hochst) benth.	Orobanchaceae	Hel	PAN	PLE ASE
186	*Rottboellia cochinchinensis* (Lour.) W. D. Clayton Syn. R. exaltata L. f.	Poaceae	Tea	PAN	BVP
187	*Saba senegalensis* (A. DC.) Pichon	Apocynaceae	Pha	AT	FVP
188	*Sapium grahamii* (Stapf) Prain	Euphorbiaceae	He m	AT	FVP
189	*Schoenefeldia gracilis* Kunth	Poaceae	Tea	AM	MVP
190	*Sclerocarya birrea* (A.Rich.) Hochst.	Anacardiaceae	Pha	AT	MVP
191	*Senna obtusifolia* (L.) H. S. Irwin & Barneby	Fabaceae	Tea	PAN	FVP

	Species	Family	TB	TP	VP
192	*Senna occidentalis* (L.) Link	Fabaceae	Tea	PAN	PLE ASE
193	*Senegalia dudgeoni* (Craib ex Holland) Kyal. & Boatwr	Fabaceae	Pha	AT	BVP
194	*Sesbania pachycarpa* DC.	Fabaceae	Pha	AT	MVP
195	*Setaria barbata* (Lam.) Kunth	Poaceae	Tea	PAN	BVP
196	*Setaria pumila* (Poiret) Roemer et Schultes	Poaceae	Pha	AT	BVP
197	*Setaria sphacelata* (Schumach.) Stapf & C.E.Hubb. Ex M.B.Moss	Poaceae	Hem	AT	BVP
198	*Sida acuta* Burm.f.	Malvaceae	Tea	PAN	FVP
199	*Sida alba* L.	Malvaceae			
200	*Sida cordifolia* L.	Malvaceae	Tea	PAN	PLE ASE
201	*Sida linifolia* Juss. ex Cavanilles	Malvaceae	Tea	PAN	FVP
202	*Sida rhombifolia* L.	Malvaceae	Hem	PAN	FVP
203	*Sida urens* L.	Malvaceae	Hem	PAN	FVP
204	*Solanum dasyphyllum* Schumach. & Thonn.	Solanaceae	Cha	AT	FVP
205	*Sorghum bicolor* (L.) Moench	Poaceae	Tea	PAN	BVP
206	*Spermacoce filifolia* J.-P. Lebrun et stork	Rubiaceae	Tea	AT	PLE ASE
207	*Spermacoce radiata* (DC.) Sieber ex Hiern	Rubiaceae	Tea	AT	PLE ASE
208	*Spermacoce ruelliae* DC.	Rubiaceae	Tea	AT	PLE ASE
209	*Spermacoce stachydea* D.C.	Rubiaceae	Tea	AT	PLE ASE
210	*Spermacoce verticillata* (Linn.) G.F.W.MEY.	Rubiaceae	Pha	PAN	PLE ASE
211	*Spigelia anthelmia* Linn.	Loganiaceae	Pha	PAN	FVP
212	*Stachytarpheta angustifolia* (Mill.) Vahl	Verbenaceae	Pha	PAN	FVP
213	*Sterculia setigera* Del.	Malvaceae	Pha	AT	MVP
214	*Striga hermonthica* (Del.) Benth	Orobanchaceae	Tea	AT	PLE ASE
215	*Strychnos innocua* Del.	Loganiaceae	Pha	PAN	PLE ASE
216	*Stylochaeton lancifolius* Kotschy. et Peyr	Araceae	Geo	AT	FVP
217	*Stylochaeton natalensis* schott	Araceae	Geo	AT	FVP
218	*Stylochaeton hypogaeus* Lepr.	Araceae	Geo	AT	FVP
219	*Stylosanthes erecta* P. Beauv.	Fabaceae	Cha	AM	BVP
220	*Synedrella nodiflora* Gaertn.	Asteraceae	Tea	PAN	FVP

	Species	Family	TB	TP	VP
221	*Tephrosia bracteolata* Guill. et Perr.	Fabaceae	Tea	AT	BVP
222	*Tephrosia linearis* (Willd.) Pers.	Fabaceae	Tea	AM	FVP
223	*Tephrosia pedicellata* Bak.	Fabaceae	Tea	AT	FVP
224	*Terminalia laxiflora* Engl. & Diels	Combretaceae	Pha	AT	PLE ASE
225	*Terminalia macroptera* Guill. & Perr.	Combretaceae	Pha	AT	PLE ASE
226	*Tribulus terrestris* L.	Zygophyllacea e	Tea	PAN	FVP
227	*Tridax procumbens* L.	Asteraceae	Tea	PAN	MVP
228	*Triumfetta rhomboidea* Jacq.	Malvaceae	Tea	PAN	PLE ASE
229	*Urena lobata* L.	Malvaceae	Tea	PAN	PLE ASE
230	*Vachellia nilotica* (L.) Willd. ex Del.	Fabaceae	Pha	PAN	BVP
231	*Vachellia sieberiana* DC.	Fabaceae	Pha	PAN	BVP
232	*Vachellia tortilis* (Forssk.) Hayne	Fabaceae	Pha	AT	BVP
233	*Vernonia cinerea* (L.) Less.	Asteraceae	Cha	PAN	FVP
234	*Vigna ambacensis* welw.wx Baker	Fabaceae	Tea	AT	MVP
235	*Vigna racemosa* (G. Don) Hutch. & Dalz.	Fabaceae	Tea	AT	MVP
236	*Vitellaria paradoxa* Gaertn.	Sapotaceae	Pha	AT	MVP
237	*Waltheria indica* L.	Malvaceae	Cha	PAN	PLE ASE
238	*Wissadula amplissima* (L.) R. E. Fries	Malvaceae	Pha	PAN	PLE ASE
239	*Zanthoxylum zanthoxyloides* (Larn.), Zepernick & Timler	Rutaceae	Pha	AT	MVP
240	*Ziziphus mauritiana* Lam.	Rhamnaceae	Pha	PAN	BVP
241	*Zornia glochidiata* C.Rchb. ex DC.	Fabaceae	Tea	AM	BVP

Biological type (TB): Chaméphyte (Cha), Géophyte (Géo), Héliophyte(Hél), Hémicryptophyte (Hém), Phanérophyte(Pha), Thérophyte(Thé).

Phytogeographic type (AT): Tropical Africa; AM: Madacascar Africa : AS-Z: SudanoZambézian Africa; AT-S: Tropical and Southern Africa; ATO: Tropical West Africa; COS: Cosmopolitan; PAL: Paleotropical; PAN: Pantropical Pastoral Value (PV): Good (G); Medium (M); Poor (L); None (W).

Appendix 4Germination monitoring sheet

Center Nationale De Semences Forestières (CNSF): Germination test
Name of observer:
Test location:............ Pretreatment:........ Date/time........... Preprocessing:.........
Sowing date:........... Sowing medium:......Species: ....
Substrate:............. Number of seeds/box:...

Box no.		Rep 1		Rep 2		Rep 3		Rep 4		Rep 5		
Observation date	Days since sowing	Tg	Tj	Tg	Tj	Tg	Tj	Tg	Tj	Tg	Tj	Infested

Nb: number of days; Rep: repeat; Tg: total germinated seeds;
Td: daily total; 1: repeat1

Appendix 5Greenhouse growth monitoring sheet

Center Nationale De Semences Forestières (CNSF): Growth monitoring								
Observer name:............ Trial location:............... Species:.........								
N° pot			Rep1	Rep2	Rep3	Rep 4	Rep 5	Status
Observation date	Days since sowing	Individual No.	H	H	H	H	H	H

H: average height
Condition: turgid, wilted, dry, dead

Appendix 6Water content determination sheet

Cup no.	P'	P"	P0 = P" - P'	Pf	Pi = Pf -P'
Species:............................. NO....................					
CNSF:............................... Date:................					
Average			P0m =		Pim =

Caption:
P' = weight of dry cups
P" = weight of dry cups + seed << wet cups >> (P" = weight of dry cups + seed << wet cups >>)
P0 = seed weight << moist >>
P0m = average weight of seeds << wet>>.
Pf = weight of dry cups + dry seeds
Pim = dry seed weight
Pim = average dry seed weight

Appendix 7List of species inventoried in the Dori communal forest

	Species	**TB**	**TP**	**VP**	**2017**	**2018**	**2019**	**2020**
1	*Acanthospermum hispidum* DC.	Pha	PAN	SVF	+	+	+	-
2	*Achyranthes aspera* L.	Cha	PAN	PLEASE	+	+	+	-
3	*Aeschynomene indica* Linn.	Tea	PAN	MVP	+	+	+	+
4	*Ageratum conyzoides* L.	Tea	PAN	PLEASE	-	+	-	+
5	*Alternanthera nodiflora* R. Br.	Cha	AT	PLEASE	+	-	-	-
6	*Alternanthera sessilis* (L.) DC.	Cha	AT	PLEASE	+	-	-	-
7	*Alysicarpus ovalifolius* (S. et Th.) Leon	Tea	PAN	BVP	+	+	+	-
8	*Alysicarpus rugosus* (Willd.) DC.	Tea	PAN	BVP	+	+	-	-
9	*Amaranthus spinosus* L.	Tea	COS	MVP	+	+	-	-
10	*Amaranthus viridis* L.	Tea	COS	MVP	+	+	-	-
11	*Andropogon gayanus* Kunth	Hem	PAN	BVP	-	-	-	+
12	*Andropogon pseudapricus* Stapf	Tea	AT	BVP	-	-	+	+
13	*Andropogon fastigiatus* sw	Tea	PAN	BVP	-	-	+	+
14	*Aristida adscensionis* L.	Tea	PAN	MVP	+	+	+	+
15	*Aristida kerstingii* Pilg.	Tea	PAN	FVP	+	+	-	+
16	*Balanites aegyptiaea* (L.) Del.	Pha	PAN	BVP	-	-	-	+
17	*Blepharis maderaspatensis* (L.) Roth	Tea	PAN	FVP	-	+	-	-
18	*Boerhavia diffusa* L.	Tea	PAN	PLEASE	+	-	-	-
19	*Brachiaria lata* (Schumach.) C. E.Hubbard	Tea	PAN	BVP	+	+	+	+
20	*Brachiaria stigmatisata* (Metz) Stapf	Tea	AT	MVP	+	+	+	+
21	*Brachiaria villosa* (Lam.) A.Camus	Tea	PAL	MVP	-	-	+	+

	Species	TB	TP	VP	2017	2018	2019	2020
22	*Bulbostylis hispidula* (Vahl) R.W.Haines	Tea	AT	PLEASE	-	-	-	-
23	*Cassia absus* L.	Tea	PAN	FVP	+	-	-	+
24	*Celosia trigyna* L.	Tea	PAN	FVP	-	-	+	-
25	*Cenchrus biflorus* Roxb.	Tea	PAN	BVP	+	+	-	-
26	*Chamaecrista mimosoides* (L.) Greene	Tea	PAL	MVP	+	+	+	-
27	*Chloris gayana* Kunth	Tea	PAN	BVP	+	+	-	-
28	*Chloris pilosa* Schum.	Tea	PAN	MVP	+	+	+	+
29	*Chrysanthellum americanum* (L.) Vatke	Tea	PAN	PLEASE	+	+	+	+
30	*Citrullus colocynthis* (L.) Schrader	Pha	AT	FVP	+	+	-	-
31	*Cleome viscosa* L.	Tea	PAN	FVP	+	-	-	-
32	*Commelina benghalensis* L.	Cha	PAN	MVP	-	-	-	-
33	*Commelina forskalaei* Vahl.	Cha	PAN	MVP	+	+	+	-
34	*Corchorus olitorius* L.	Tea	PAN	MVP	+	+	+	-
35	*Corchorus tridens* L.	Tea	PAN	MVP	+	-	-	-
36	*Cyanotis lanata* Benth.	Tea	AT	BVP	+	-	-	-
37	*Cyperus difformis* Linn.	Tea	PAN	FVP	+	-	-	+
38	*Cyperus dilatatus* Schumach. & Thonn.	Geo	AT	FVP	+	+	+	+
39	*Cyperus esculentus* Linn.	Geo	PAN	PLEASE	+	+	+	+
40	*Cyperus iria* Linn.	Tea	PAN	FVP	+	+	+	+
41	*Cyperus rotundus* L.	Geo	PAN	FVP	+	+	+	-
42	*Cyperus sphacelatus* Rottb.	Hem	PAN	FVP	-	-	+	-
43	*Dactyloctenium aegyptium* (Linn) Wild.	Tea	PAL	BVP	+	+	+	+
44	*Desmodium hirtum* Guill & Perr.	Tea	AT	MVP	+	+	-	+

	Species	TB	TP	VP	2017	2018	2019	2020
45	*Dichrostachys cinerea* L.Wight & Arn.	Pha	AM	BVP	+	+	+	+
46	*Digitaria horizontalis* Willd.	Tea	PAN	MVP	+	+	+	+
47	*Echinochloa colona* (L.) Link	Tea	PAN	MVP	+	+	+	-
48	*Echinochloa stagnina* (Retz.) P.Beauv.	Tea	PAN	MVP	+	+	+	-
49	*Elionurus elegans* Kunth	Tea	AT	MVP	+	+	+	-
50	*Entada africana* Guill. et Perr.	Pha	AM	BVP	-	-	+	+
51	Eragrostris cilianensis L.	Cha	AT	MVP	+	+	-	+
52	*Eragrostis ciliaris* (L.) R. Br.	Tea	PAN	MVP	+	+	-	-
53	*Eragrostis tenella* (L.) Roem. & Schult.	Tea	AT	MVP	+	+	-	-
54	*Eragrostis tremula* Hochst.	Tea	AT	MVP	+	+	+	+
55	*Euphorbia heterophylla* L.	Tea	PAN	MVP	-	+	-	-
56	*Euphorbia hirta* L.	Tea	PAN	MVP	+	-	-	-
57	*Evolvolus alsinoides* (Linn.)	Tea	AT	FVP	+	+	-	-
58	*Fimbristylis ferurginea* (Linn.) Vahl	Pha	PAN	FVP	+	+	+	-
59	*Hackelochloa granularis* (L.) Kuntze	Tea	PAN	MVP	+	+	+	-
60	*Hibiscus cannabinus* L.	Tea	PAN	MVP	-	+	+	-
61	*Hygrophila auriculata* (Schumach.) Heine	Tea	PAN	FVP	-	+	-	-
62	*Indigofera bracteolata* DC.	Cha	AT	MVP	+	-	-	-
63	*Indigofera hirsuta* L.	Tea	PAN	MVP	+	+	+	-
64	*Indigofera stenophylla* Guill. & Perr.	Tea	AT	FVP	+	+	+	+
65	*Indigofera trichopoda* Lepr. ex Guill.& Perr.	Tea	AT	PLEASE	+	+	+	+
66	*Ipomoea aquatica* Forsk.	Tea	PAN	FVP	-	-	+	-
67	*Ipomoea argentaurata* Hall. f.	Tea	AT	MVP	+	+	+	-

	Species	TB	TP	VP	2017	2018	2019	2020
68	*Ipomoea asarifolia* (Desr.) Roem. & Schult.	Tea	AT	MVP	-	-	-	+
69	*Ipomoea coscinosperma* Hochst. ex Choisy	Tea	AT	BVP	+	+	-	+
70	*Ipomoea eriocarpa* R. Br.	Tea	PAL	BVP	+	+	-	-
71	*Ipomea triloba* L.	Tea	PAN	BVP	-	-	+	+
72	*Kyllinga erecta* Schumach.	Geo	PAN	PLEASE	+	+	+	+
73	*Kyllinga pumila* Michx	Hem	PAN	PLEASE	+	+	+	+
74	*Kyllinga squamulata* Thonn. ex Vahl	Tea	PAN	PLEASE	+	+	+	+
75	*Leptadenia hastata* (Pers.) Decne	Pha	AT	FVP	+	+	-	-
76	*Leucas martinicensis* (Jacq.) R. Br.	Tea	PAN	PLEASE	+	+	+	-
77	*Loudetia togoensis (Pilg).* C.E Hubbard	Tea	AT	FVP	-	+	-	-
78	*Ludwigia abyssinica* A. Rich.	Pha	AT	PLEASE	-	-	-	+
79	*Ludwigia hyssopifolia* (G. Don.) Exell	Tea	PAN	PLEASE	-	-	-	+
80	*Marsilea minuta* L.	Hel	PAN	PLEASE	+	+	+	-
81	*Melochia corchorifolia* Linn.	Tea	PAL	MVP	+	-	-	-
82	*Microchloa indica* (L.f.) P.Beauv.	Tea	PAN	PLEASE	+	-	-	-
83	*Mollugo nudicaulis* Lam.	Tea	PAN	PLEASE	+	+	-	-
84	*Oldenlandia corymbosa* Linn.	Tea	PAN	MVP	+	-	+	-
85	*Panicum laetum* Kunth	Tea	AT	MVP	+	+	+	-
86	*Pennisetum pedicellatum* Trin.	Tea	PAN	BVP	+	+	+	+
87	*Pennicetum polystachion* (Linnaeus) Schultes				-	-	-	+
88	*Peristrophe bicalyculata* (Retz.) Nees	Tea	PAN	FVP	-	-	-	+
89	*Phyllanthus amarus* Sch. Et Th.	Tea	PAN	FVP	+	-	-	-
90	*Physalis angulata* L.	Tea	PAN	PLEASE	+	-	-	+

	Species	TB	TP	VP	2017	2018	2019	2020
91	*Polycarpaea corymbosa* (L) Lam.	Tea	PAN	PLE ASE	+	-	-	-
92	*Polycarpaea linearifolia* Lam	Tea	AT	PLE ASE	+	+	-	-
93	*Polygala multiflora* Poiret	Tea	AT	PLE ASE	+	-	-	-
94	*Portulaca oleracea* L.	Tea	COS	MVP	+	+	-	-
95	*Prosopis juliflora* (Sw.) DC.	Pha	PAN	MVP	-	-	+	+
96	*Pycreus flavescens* (L.) P.Beauv. ex Rchb.	Hel	PAN	FVP	+	+	-	+
97	*Ramphicarpa fistulosa* (Hochst) benth.	Hel	PAN	PLE ASE	+	+	-	-
98	*Schizachyrium ruderale* Clayton.	Tea	AT	BVP	-	+	+	+
99	*Schoenefeldia gracilis* Kunth	Tea	AM	MVP	+	+	+	+
100	*Senna obtusifolia* (L.) H. S. Irwin & Barneby	Tea	PAN	FVP	+	+	+	+
101	*Senna occidentalis* (L.) Link	Tea	PAN	PLE ASE	+	+	-	-
102	*Senegalia dudgeoni* (Craib ex Holland) Kyal. & Boatwr	Pha	AT	BVP	+	+	+	+
103	*Setaria barbata* (Lam.) Kunth	Tea	PAN	BVP	+	+	+	+
104	*Setaria pumila* (Poiret) Roemer et Schultes	Pha	AT	BVP	+	+	+	+
105	*Setaria verticillata*(Linnaeus) Palisot de beauvois	Tea	AT	BVP	+	+	+	+
106	*Sida alba* L.	Tea	PAN	FVP	+	-	-	-
107	*Sida cordifolia* L.	Tea	PAN	PLE ASE	+	-	-	-
108	*Sida urens* L.	Hem	PAN	FVP	+	+	-	+
109	*Sorghum bicolor* (L.) Moench	Tea	PAN	BVP	+	-	-	-
110	*Spermacoce chaetocephala* **DC.**	Tea	AM	PLE ASE	+	+	+	+
111	*Spermacoce filifolia* J.-P. Lebrun et stork	Tea	AT	PLE ASE	+	+	+	+
112	*Spermacoce radiata* (DC.) Sieber ex Hiern	Tea	AT	PLE ASE	+	+	-	+

	Species	**TB**	**TP**	**VP**	**2017**	**2018**	**2019**	**2020**
113	*Spermacoce ruelliae* DC.	Tea	AT	PLE ASE	+	-	-	+
114	*Spermacoce stachydea* D.C.	Tea	AT	PLE ASE	+	+	+	+
115	*Spermacoce verticillata* (Linn.) G.F.W.MEY.	Pha	PAN	PLE ASE	-	+	+	+
116	*Striga hermonthica* (Del.) Benth	Tea	AT	PLE ASE	-	+	+	-
117	*Stylosanthes erecta* P. Beauv.	Cha	AM	BVP	+	+	+	+
118	*Tephrosia bracteolata* Guill. et Perr.	Tea	AT	BVP	+	+	+	-
119	*Tephrosia pedicellata* Bak.	Tea	AT	FVP	+	-	+	-
120	*Tribulus terrestris* L.	Tea	PAN	FVP	+	+	+	+
121	*Tridax procumbens* L.	Tea	PAN	MVP	+	-	-	-
122	*Triumfetta rhomboidea* Jacq.	Tea	PAN	PLE ASE	+	-	-	-
123	*Vachellia nilotica* (L.) Willd. ex Del.	Pha	PAN	BVP	-	-	+	+
124	*Vachellia sieberiana* DC.	Pha	PAN	BVP	-	-	+	+
125	*Vachellia seyal* (Del.) P.J.H.Hurter,	Pha	AT	BVP	-	-	+	+
126	*Vachellia tortilis* (Forssk.) Hayne	Pha	AT	BVP	-	-	+	+
127	*Waltheria indica* L.	Cha	PAN	PLE ASE	+	+		
128	*Wissadula amplissima* (L.) R. E. Fries	Pha	PAN	PLE ASE	+	-	-	-
129	*Ziziphus mauritiana* Lam.	Pha	PAN	BVP	-	-	+	+
130	*Zornia glochidiata* C.Rchb. ex DC.	Tea	AM	BVP	+	+	+	+

+ indicates the presence of the species in the environment; - indicates the absence of the species in the environment. Biological type (TB): Chaméphyte (Cha), Géophyte (Géo), Héliophyte(Hél), Hémicryptophyte (Hém), Phanérophyte(Pha), Thérophyte(Thé).
Phytogeographical type (AT): Tropical Africa; AM: Madacascar Africa : AS-Z: SudanoZambézian Africa; AT-S: Tropical and Southern Africa; ATO: Tropical West Africa; COS: Cosmopolitan; PAL: Paleotropical; PAN: Pantropical
Pastoral Value (PV): Good (G); Medium (M); Poor (L); None (W).

Printed by Books on Demand GmbH, Norderstedt / Germany